Grade 6

ughton
flin
ourt

Getting Ready for
High-Stakes
Assessments

INCLUDES:

- Standards Practice
- Beginning-, Middle-, and End-of-Year Benchmark Tests
  with Performance Tasks
- Year-End Performance Assessment Task

Printed in the U.S.A.

ISBN 978-0-544-60197-0

3 4 5 6 7   0304   24 23 22 21 20

4500792631   A B C D E F G

Contents

Contents

High-Stakes Assessment Item Formats

The various high-stakes assessments contain item types beyond the traditional multiple-choice format, which allows for a more robust assessment of students' understanding of concepts.

The high-stakes mathematics assessments for Grade 6 allows for calculator use for certain portions of the test. During these calculator sessions in the Grade 6 test, a four-function calculator, with square root function, will be allowed.

In this book, items for which calculator use is allowed are identified with a .

Most high-stakes assessments will be administered via computers; and *Getting Ready for High-Stakes Assessments* presents items in formats similar to what you will see on the tests. The following information is provided to help familiarize you with these different types of items. Each item type is identified on pages (vii–viii). The examples will introduce you to the item types.

The following explanations are provided to guide you in answering the questions. These pages (v–vi) describe the most common item types. You may find other types on some tests.

Example 1 Identify examples of a property.

More Than One Correct Choice

This type of item looks like a traditional multiple-choice item, but it asks you to mark all that apply. To mark all that apply, look for more than one correct choice. Carefully look at each choice and mark it if it is a correct answer.

Example 2 Circle the word that completes the sentence.

Choose From a List

Sometimes when you take a test on a computer, you will have to select a word, number, or symbol from a drop-down list. The *Getting Ready for High-Stakes Assessments* tests show a list and ask you to choose the correct answer. Make your choice by circling the correct answer. There will only be one choice that is correct.

Example 3 Sort numbers by categories for multiples.

Sorting

You may be asked to sort something into categories. These items will present numbers, words, or equations on rectangular "tiles." The directions will ask you to write each of the items in the box that describes it.

When the sorting involves more complex equations or drawings, each tile will have a letter next to it. You will be asked to write the letter for the tile in the box. Sometimes you may write the same number or word in more than one box. For example, if you need to sort quadrilaterals by category, a square could be in a box labeled *rectangle* and another box labeled *rhombus*.

Example 4 Order numbers from least to greatest.

Use Given Numbers in the Answer

You may also see numbers and symbols on tiles when you are asked to write an equation or answer a question using only numbers. You should use the given numbers to write the answer to the problem. Sometimes there will be extra numbers. You may need to use each number more than once.

Example 5 Match related facts.

Matching

Some items will ask you to match equivalent values or other related items. The directions will specify what you should match. There will be dots to guide you in drawing lines. The matching may be between columns or rows.

Item Types:

Example 1

More Than One
Correct Choice

Fill in the bubble
next to all the correct
answers.

Select the equations that show the Commutative
Property of Multiplication. Mark all that apply.

(A) $35 \times 56 = (30 + 5) \times (50 + 6)$

(B) $47 \times 68 = 68 \times 47$

(C) $32 \times 54 = 54 \times 32$

(D) $12 \times 90 = 90 \times 12$

(E) $34 \times 932 = 34 \times (900 + 30 + 2)$

(F) $45 \times 167 = (40 + 5) \times 167$

Example 2

Choose From a List

Circle the word that
completes the sentence.

$(25 \times 17) \times 20 = 25 \times (17 \times 20)$

The equation shows
the factors in a different

order.
grouping.
operation.

Example 3

Sorting

Copy the numbers in the correct box.

Write each number in the box below the term that describes it.

| 30 | 42 | 72 | 85 |

Multiple of 5	Multiple of 6

Example 4

Use Given Numbers in the Answer

Write the given numbers to answer the question.

Write the numbers in order from least to greatest.

| 18,345 | 17,467 | 18,714 | 16,235 |

_____

Example 5

Matching

Draw lines to match an item in one column to the related item in the other column.

Match the pairs of related facts.

8 × 7 = 56   •   •   8 × 9 = 72

8 × 6 = 48   •   •   7 × 8 = 56

72 ÷ 9 = 8   •   •   9 × 7 = 63

63 ÷ 7 = 9   •   •   48 ÷ 6 = 8

1. Kendra has 4 necklaces, 7 bracelets, and 5 rings. Draw a model to show the ratio that compares rings to bracelets.

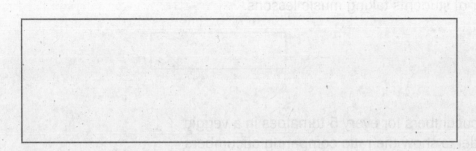

2. There are 3 girls and 2 boys taking swimming lessons. Write the ratio that compares the number of girls taking swimming lessons to the total number of students taking swimming lessons.

_____

3. Luis adds 3 strawberries for every 2 blueberries in his fruit smoothie. Draw a model to show the ratio that compares the number of strawberries to the number of blueberries.

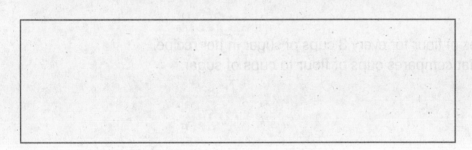

4. Sam has 3 green apples and 4 red apples. Select the ratios that compare the number of red apples to the total number of apples. Mark all that apply.

(A) 4 to 7   (D) 4 : 3

(B) 3 to 7   (E) $\frac{3}{7}$

(C) 4 : 7   (F) $\frac{4}{7}$

GO ON ➤

5. There are 3 girls and 4 boys taking music lessons. Write the
ratio that compares the number of boys taking music lessons to
the total number of students taking music lessons.

6. Camilla adds 2 cucumbers for every 5 tomatoes in a veggie
mix. Draw a model to show the ratio comparing cucumbers
to tomatoes.

7. Write the ratio 4 to 9 in two different ways.

_____

8. Zena adds 4 cups of flour for every 3 cups of sugar in her recipe.
Draw a model that compares cups of flour to cups of sugar.

9. Julia has 2 green reusable shopping bags and 5 purple
reusable shopping bags. Select the ratios that compare the
number of purple reusable shopping bags to the total number
of reusable shopping bags. Mark all that apply.

(A) 5 to 7   (D) 5 : 2

(B) 5 : 7   (E) $\frac{2}{5}$

(C) 2 to 7   (F) $\frac{5}{7}$

Name _____

Practice Test
6.RP.A.2
Understand ratio concepts and use
ratio reasoning to solve problems.

1. A company called Your Home Builders advertises that it can build a new home at a price of $390 for every 15 square feet. Another company called Fast Right Now charges $330 for every 11 square feet to build a new home. Which company charges less per square foot? Use numbers and words to explain your answer.

2. Abby goes to the pool to swim laps. The graph shows how far Abby swam over time. Use equivalent ratios to find how far Abby swam in 7 minutes.

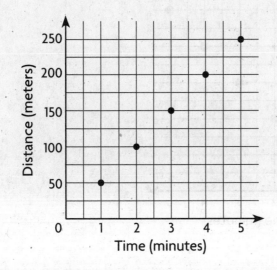

Time (minutes)

_____ meters

3. A rabbit runs 35 miles per hour. Select the animals who run at a faster unit rate per hour than the rabbit. Mark all that apply.

Ⓐ Reindeer: 100 miles in 2 hours

Ⓑ Ostrich: 80 miles in 2 hours

Ⓒ Zebra: 90 miles in 3 hours

Ⓓ Squirrel: 36 miles in 3 hours

GO ON ➡

4. A construction company pays its workers $288 for 12 hours of work. A demolition company pays its workers $225 for 9 hours of work. Which company pays less per hour? Use numbers and words to explain your answer.

5. Marc enjoys running. The graph shows how far Marc ran over time. Use equivalent ratios to find how far Marc ran in 7 minutes.

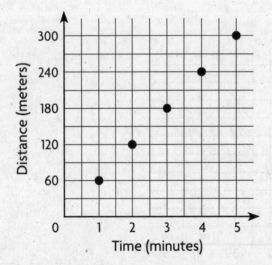

_____ meters

6. An insurance company offers a rate of $119 per month for a health plan. Select the companies that offer a plan with a lower unit rate. Mark all that apply.

Ⓐ Company A: $250 for 2 months

Ⓑ Company B: $348 for 3 months

Ⓒ Company C: $380 for 4 months

Ⓓ Company D: $500 for 4 months

Name _____

1. Jeff ran 2 miles in 12 minutes. Ju Chan ran 3 miles in 18 minutes. Did Jeff and Ju Chan run the same number of miles per minute? Complete the tables of equivalent ratios to support your answer.

Jeff				
Distance (miles)	2			
Time (minutes)	12			

Ju Chan				
Distance (miles)	3			
Time (minutes)	18			

2. Water is filling a bathtub at a rate of 3 gallons per minute.

Part A

Complete the table of equivalent ratios for the first 5 minutes of the bathtub filling up.

Amount of Water (gallons)	3				
Time (minutes)	1				

Part B

Emily said there will be 36 gallons of water in the bathtub after 12 minutes. Explain how Emily could have found her answer.

3. Look at the numbers on the tiles. Determine whether each ratio is equivalent to $\frac{1}{2}$, $\frac{3}{9}$, or $\frac{5}{6}$. Write the ratio in the correct box.

| $\frac{2}{6}$ | $\frac{3}{6}$ | $\frac{5}{10}$ | $\frac{10}{14}$ | $\frac{50}{100}$ | $\frac{20}{28}$ | $\frac{1}{3}$ | $\frac{8}{24}$ |

$\frac{1}{2}$	$\frac{3}{9}$	$\frac{5}{7}$

4. Edgar said $\frac{3}{5}$ is equivalent to $\frac{18}{32}$. Check his work by completing the table of equivalent ratios. Is Edgar correct? Explain your answer.

3				
5				

_____

5. The Garcias are driving to the beach. They are traveling at a rate of 30 miles per hour. Use the ordered pairs to graph the distance traveled over time.

Distance (miles)	30	60	90	120	150
Time (hours)	1	2	3	4	5

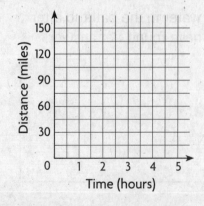

1. Scotty earns $35 for babysitting for 5 hours. If Scotty always charges the same rate, how many hours will it take him to earn $42?

_____ hours

2. Caleb bought 6 packs of pencils for $12.

Part A

How much will he pay for 9 packs of pencils? Use numbers and words to explain your answer.

Part B

Describe how to use a bar model to solve the problem.

3. Peri earned $27 for walking her neighbor's dog 3 times. If Peri charged the same rate and earned $36, how many times did she walk her neighbor's dog? Use a unit rate to find the unknown value.

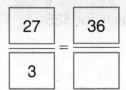

_____ times

4. Match each situation to its unit rate.

9 boxes for $54	6 bags for $42	12 tablets for $24	4 pounds for $12
•	•	•	•
•	•	•	•
1 to 2	1 to 3	1 to 6	1 to 7

GO ON

Name _____

5. Vicki earns $30 for washing 6 cars. If Vicki always charges the same rate, how many hours will it take her to earn $35?

_____ hours

6. Kayden bought 9 packs of paper for $27.

Part A

How much will he pay for 11 packs of paper? Use numbers and words to explain your answer.

Part B

Describe how to use a bar model to solve the problem.

7. Match each situation to its unit rate.

12 pages for $72   6 ounces for $12   8 bags for $24   4 cases for $28

•   •   •   •

•   •   •   •

1 to 2   1 to 3   1 to 6   1 to 7

8. Melinda rides her bike 18 miles in 2 hours. If she rides at a constant speed, select the answers below that are equivalent ratios to the speed at which she rides. Select all ratios that are equivalent.

(A) 27 miles in 4 hours

(B) 9 miles in 1 hour

(C) 36 miles in 2 hours

(D) 27 miles in 3 hours

(E) 36 miles in 4 hours

1. In which pair of values do the percent and the fraction represent the same amount?

Ⓐ 12% and $\frac{1}{2}$

Ⓑ 45% and $\frac{4}{5}$

Ⓒ $\frac{3}{8}$ and 37.5%

Ⓓ $\frac{2}{10}$ and 210%

 2. The school orchestra has 25 woodwinds, 15 percussionists, 30 strings, and 30 brass instruments. Select the portion of the instruments that are percussion. Mark all that apply.

Ⓐ 15%

Ⓑ 1.5

Ⓒ $\frac{3}{20}$

Ⓓ 0.15

 3. For a science project, $\frac{3}{4}$ of the students chose to make a poster and 0.25 of the students wrote a report. Rosa said that more students made a poster than wrote a report. Do you agree with Rosa? Use numbers and words to support your answer.

4. There are 22 red marbles in a bin of marbles. Red marbles make up 25% of all the marbles.

There are
| 3 |
| 78 |
| 88 |
| 100 |
total marbles in the bin.

GO ON

Name _____

5. For numbers 5a–5b, choose <, >, or =.

5a. 25% of 60
$$< \\ > \\ =$$
40% of 30

5b. 30% of 60
$$< \\ > \\ =$$
75% of 40

6. Avery wants to put a variety of muffins in a display case. The case is large enough to hold 60 muffins.

Part A

Complete the table.

Type of Muffin	Percent of Maximum Number	Number of Muffins in Case
Blueberry	40%	
Pumpkin	20%	
Cranberry	30%	

Part B

Has Avery put the maximum number of muffins in the case? Use numbers and words to explain how you know. If she has not put the maximum number in the case, how many more muffins could she put in the case?

Practice Test

6.RP.A.3d
*Understand ratio concepts and use
ratio reasoning to solve problems.*

1. A construction crew needs to remove 6 tons of river rock during the construction of new office buildings.

The weight of the rocks is

| 1,200 |
| 6,000 |
| 12,000 |
| 18,000 |

pounds.

2. Select the conversions that are equivalent to 10 yards. Mark all that apply.

(A) 20 feet   (C) 30 feet

(B) 240 inches   (D) 360 inches

3. Harry received a package for his birthday. The package weighed 357,000 centigrams. Select the conversions that are equivalent to 357,000 centigrams. Mark all that apply.

(A) 3.57 kilograms

(B) 357 dekagrams

(C) 3,570 grams

(D) 3,570,000 decigrams

4. Nadia has a can of vegetables with a mass of 411 grams. Write equivalent conversions for 411 grams in the correct boxes.

| 4.11 | 41.1 | 0.411 |

kilograms	hectograms	dekagrams

GO ON

5. Select the conversions that are equivalent to 25 yards.
Mark all that apply.

(A) 50 feet (B) 75 feet

(C) 900 inches (D) 1,000 inches

6. A rectangular room measures 14 feet by 144 inches. Edgar said
the area of the room is 2,016 square feet. Explain his mistake,
and then find the area in square feet.

7. Claire says that if she runs at an average rate of 6 miles per
hour, it will take her about 2 hours to run 18 miles. Do you
agree or disagree with Claire? Use numbers and words to
support your answer.

_____

_____

_____

_____

8. The Wilson family's newborn baby weighs 84 ounces.
Choose the numbers to show the baby's weight in
pounds and ounces.

| 5 |
| 6 | pounds
| 7 |

| 3 |
| 4 | ounces
| 5 |

Practice Test

6.NS.A.1
Apply and extend previous understandings of multiplication and division to divide fractions by fractions.

1. Complete the table by finding the products. Then answer the questions in Part A and Part B.

Division	Multiplication
$\dfrac{1}{5} \div \dfrac{3}{4} = \dfrac{4}{15}$	$\dfrac{1}{5} \times \dfrac{4}{3} =$
$\dfrac{2}{13} \div \dfrac{1}{5} = \dfrac{10}{13}$	$\dfrac{2}{13} \times \dfrac{5}{1} =$
$\dfrac{4}{5} \div \dfrac{3}{5} = \dfrac{4}{3}$	$\dfrac{4}{5} \times \dfrac{5}{3} =$

Part A

Explain how each pair of division and multiplication problems are the same and how they are different.

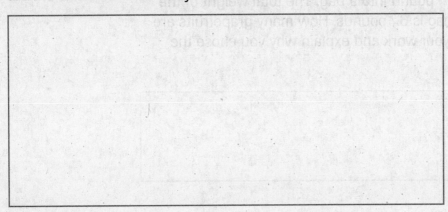

Part B

Explain how to use the pattern in the table to rewrite a division problem involving fractions as a multiplication problem.

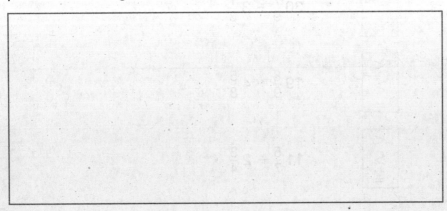

GO ON

2. Tricia has a bag of grapes weighing $1\frac{1}{4}$ pounds. She wants to divide the grapes into equal shares that will each weigh $\frac{5}{16}$ pound. How many equal shares can she create?

Ⓐ 1

Ⓑ 2

Ⓒ 4

Ⓓ 5

3. $\frac{6}{7} \div \frac{1}{3} = \boxed{}$

4. Jillian picks some grapefruits. She places all the grapefruits that weigh exactly $\frac{1}{2}$ pound into a bag. The total weight of the grapefruits in the bag is $6\frac{1}{2}$ pounds. How many grapefruits are in the bag? Show your work and explain why you chose the operation you did.

5. For numbers 5a–5c, estimate to compare. Choose $<$, $>$, or $=$.

5a. $18\frac{3}{10} \div 2\frac{5}{6}$ $30\frac{7}{9} \div 3\frac{1}{3}$

5b. $17\frac{4}{5} \div 6\frac{1}{6}$ $\begin{array}{c}<\\>\\=\end{array}$ $19\frac{8}{9} \div 4\frac{5}{8}$

5c. $35\frac{5}{6} \div 6\frac{1}{4}$ $\begin{array}{c}<\\>\\=\end{array}$ $11\frac{5}{7} \div 2\frac{3}{4}$

1. Select the equations that are correct. Mark all that apply.

Ⓐ 17,550 ÷ 65 = 270

Ⓑ 11,196 ÷ 12 = 933

Ⓒ 29,365 ÷ 35 = 677

Ⓓ 11,712 ÷ 96 = 122

2. For numbers 2a–2d, write the number that completes the equation.

2a. $17,358 \div \boxed{} = 789$

2b. $\boxed{} \div 68 = 378$

2c. $58,113 \div 99 = \boxed{}$

2d. $53,680 \div 88 = \boxed{}$

3. An organism measures 11,400 millimeters in length in a photograph. If the photo has been enlarged by a factor of 75, what is the actual length of the organism? Show your work.

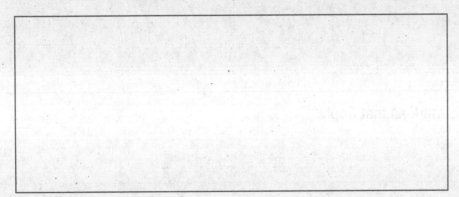

GO ON

Name _____

4. Select the equations that are correct. Mark all that apply.

Ⓐ $14,364 \div 63 = 228$

Ⓑ $32,536 \div 49 = 664$

Ⓒ $50,031 \div 51 = 981$

Ⓓ $44,280 \div 82 = 987$

5. Derek uses 10,128 wooden sticks in a model of a tower. If wooden sticks come in packages of 16, how many packages were needed to make the model? Show your work.

6. Select the quotient. Mark all that apply.
$24\overline{)13,206}$

Ⓐ   550.12

Ⓑ   550 R6

Ⓒ   550.25

Ⓓ   550 R24

7. Select the quotient. Mark all that apply.
$40\overline{)21,488}$

Ⓐ   519 R6

Ⓑ   537 R2

Ⓒ   537 R8

Ⓓ   537.2

Name _____

1. Francesca's sunflower was 1.75 feet tall at the end of April. It grew 4.392 feet from then until the end of July. How tall was her sunflower at the end of July?

_____ feet

2. Carlita went shopping at an art supply store. The table shows the weight of each type of clay she bought and the total cost. Complete the table to show the cost per pound of each type of clay.

Type of Clay	Weight of Clay Purchased (in pounds)	Total Cost (in dollars)	Cost per Pound (in dollars)
Plasticlay	1.68	$6.72	
Ceramic Clay	1.25	$5.50	
Polymer Clay	3.48	$5.22	

3. The distance around the outside of Cedar Park is 11.8 miles. Joanie ran 0.25 of the distance during her lunch break. How far did she run? Show your work.

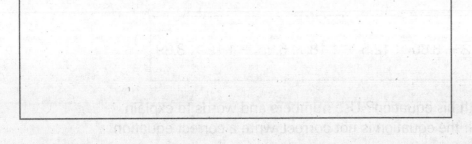

GO ON

Name _____

4. The Lowe family is going to a swim meet. They buy 12.5 liters of juice for $1.18 per liter, 6.25 pounds of ice for $1.12 per pound, and a bag of oranges for $8.89. Before they leave, they fill up the car with 10.2 gallons of gasoline at a cost of $3.80 per gallon.

Part A

Complete the table by calculating the total cost for each item.

Item	Calculation	Total Cost
Gasoline	10.2 × 3.80	
Juice	12.5 × 1.18	
Ice	6.25 × 1.12	
Oranges	8.89	

Part B

What is the total cost for everything before tax? Show your work.

Part C

Mr. Lowe calculates the total cost for everything before tax using this equation.

Total cost = 10.2 + 3.80 × 12.5 + 1.18 × 6.25 + 1.12 × 8.89

Do you agree with his equation? Use numbers and words to explain why or why not. If the equation is not correct, write a correct equation.

1. Select two numbers that have 9 as their greatest common factor. Mark all that apply.

(A) 3, 9

(B) 3, 18

(C) 9, 18

(D) 9, 36

(E) 18, 27

2. The prime factorization of each number is shown.

$$15 = 3 \times 5$$
$$18 = 2 \times 3 \times 3$$

Part A

Using the prime factorization, complete the Venn diagram.

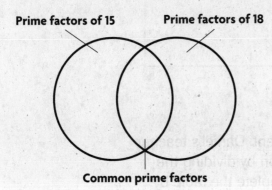

Prime factors of 15   Prime factors of 18

Common prime factors

Part B

Find the GCF of 15 and 18.

_____

3. Two-fifths of the fish in Gary's fish tank are guppies. One-fourth of the guppies are red. What fraction of the fish in Gary's tank are red guppies? Show your work.

GO ON

4. There are 20 sixth graders and 25 seventh graders in the Junior Scholar Club. For their first community service project, the Scholar Club president wants to organize the club members into equal-sized groups. Each group will have only sixth graders or only seventh graders.

Part A

How many students will be in each group if each group has the greatest possible number of members? Show your work.

Part B

If each group has the greatest possible number of club members, how many groups of sixth graders and how many groups of seventh graders will there be? Use numbers and words to explain your answer.

5. The table shows Daniel's homework assignment. Daniel's teacher instructed the class to simplify each expression by dividing the numerator and denominator by the GCF. Complete the table by simplifying each expression and then finding the product.

Problem	Expression	Simplified Expression	Product
a	$\frac{5}{9} \times \frac{3}{10}$		
b	$\frac{3}{5} \times \frac{2}{7}$		
c	$\frac{5}{7} \times \frac{7}{10}$		
d	$\frac{4}{5} \times \frac{1}{8}$		

Practice Test

6.NS.C.5
Apply and extend previous understandings of numbers to the system of rational numbers.

1. Select the situations that can be represented by a negative number. Mark all that apply.

   (A) Death Valley is located 282 feet below sea level.

   (B) Austin's golf score was 3 strokes below par.

   (C) The average temperature in Santa Monica in August is 75°F.

   (D) Janai withdraws $20 from her bank account.

2. Select the situations that can be represented by a negative number. Mark all that apply.

   (A) Sherri lost 100 points answering a question wrong.

   (B) The peak of a mountain is 2,000 feet above sea level.

   (C) Yong paid $25 for a parking ticket.

   (D) A puppy gained 3 pounds.

3. Bombay Beach in California is 225 feet below sea level. What integer can be used to express this depth?

   _____

4. Which situation could be represented by the integer ⁻4?

   (A) A football team gains 4 yards on a play.

   (B) A student answers a 4-point question correctly.

   (C) A temperature is 4°F below zero.

   (D) An elevation is 4 feet above sea level.

GO ON

Name _____

5. Select the situations that could be represented by the integer ⁺3. Mark all that apply.

Ⓐ A football team gains 3 yards on a play.

Ⓑ A golfer's score is 3 over par.

Ⓒ A student answers a 3-point question correctly.

Ⓓ A cat loses 3 pounds.

6. Mr. Williams went scuba diving and took photographs of sea life at 25 feet below sea level. If sea level is represented by 0, write an integer to represent the depth at which Mr. Williams took photographs. Explain your answer.

7. Joyce recorded Tuesday's temperature as ⁻5°F. Describe where this number appears on a thermometer. Explain what ⁻5 means in terms of temperature.

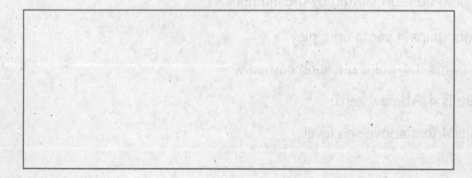

1. A flag pole is located at point 0 on a map of Orange Avenue. Other points of interest on Orange Avenue are indicated by their distances, in miles, to the right of the flag pole (positive numbers) or to the left of the flag pole (negative numbers). Graph and label each location on the number line.

Name	Location
School	0.4
Post Office	1.8
Library	⁻1
Fire Station	⁻1.3

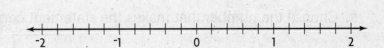

2. Select the numbers that are between ⁻1 and ⁻2. Mark all that apply.

(A) $\frac{-4}{5}$

(B) $1\frac{2}{3}$

(C) ⁻1.3

(D) $^-1\frac{1}{4}$

(E) $^-2\frac{1}{10}$

3. Choose the number that makes the statement correct.

On a number line, $^-3\frac{5}{8}$ is between ⁻3 and 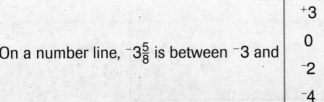 .

GO ON

Name _____

4. Select the numbers that are between ⁻1 and 1. Mark all that apply.

Ⓐ $-\frac{4}{5}$

Ⓑ ⁻0.9

Ⓒ $1\frac{1}{4}$

Ⓓ $-1\frac{1}{10}$

5. Choose the number that makes the statement correct.

On a number line, $1\frac{7}{8}$ is between 1 and

⁻1
0
+7
8
+2

6. A thermometer shows a temperature of ⁻4.5°C. A nearby thermometer shows a temperature of ⁻3.5°C. Explain how absolute value can be used to decide which temperature is warmer.

Name _____

Practice Test

6.NS.C.6b
Apply and extend previous understandings of numbers to the system of rational numbers.

1. Identify the quadrant where each point is located. Write each point in the correct box.

   (⁻1, 3)   (4, ⁻2)   (⁻3, ⁻2)

   (1, ⁻3)   (⁻1, 2)   (3, 4)

Quadrant I	Quadrant II	Quadrant III	Quadrant IV

2. Mia's house is located at point (3, 4) on a coordinate plane. The location of Keisha's house is the reflection of Mia's house across the y-axis. In what quadrant is Keisha's house?

   _____

3. Point A (2, ⁻3) is reflected across the x-axis to point B. Point B is reflected across the y-axis to point C. What are the coordinates of point C? Use words and numbers to explain your answer.

Name _____

4. Rex's house is located at point (2, ⁻5) on a coordinate plane. The location of Terrell's house is the reflection of the coordinates of Rex's house across the *x*-axis. In what quadrant is Terrell's house?

_____

5. Point *R* (4, ⁻5) is reflected across the *y*-axis to point *S*. Point *S* is reflected across the *x*-axis to point *T*. What are the coordinates of point *T*? Use words and numbers to explain your answer.

6. Identify the quadrant where each point is located. Write each point in the correct box.

(⁻5, 2)   (6, ⁻4)   (⁻1, ⁻9)

(5, ⁻4)   (⁻3, 3)   (7, 2)

Quadrant I	Quadrant II	Quadrant III	Quadrant IV

Name _____

Practice Test

6.NS.C.6c
Apply and extend previous understandings of numbers to the system of rational numbers.

1. Which point on the number line represents $\frac{4}{5}$?

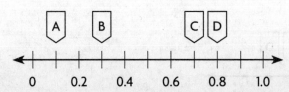

Ⓐ Point A

Ⓑ Point B

Ⓒ Point C

Ⓓ Point D

2. Choose the word that makes the statement correct.

If both the x- and y-coordinates are
| positive |
| negative |
| equal |
, the point is

always to the left of the y-axis and below the x-axis.

3. Explain how to graph points $A(^-3, 0)$, $B(0, 0)$, and $C(0, ^-3)$ on the coordinate plane. Then, explain how to graph point D so that ABCD is a square.

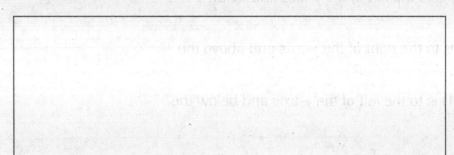

GO ON ▶

Name _____

4. Write the decimal and fraction in simplest form represented by each point.

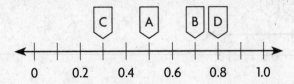

Point A ☐   Point B ☐

Point C ☐   Point D ☐

5. Write the values in order from least to greatest.

| $\frac{1}{3}$ | 0.45 | 0.39 | $\frac{2}{5}$ |

_____ _____ _____ _____

6. Select the statements that are correct. Mark all that apply.

Ⓐ Point A (2, ⁻1) is to the right of the y-axis and below the x-axis.

Ⓑ Point B (⁻5, 2) is to the left of the y-axis and below the x-axis.

Ⓒ Point C (3, 2) is to the right of the y-axis and above the x-axis.

Ⓓ Point D (⁻2, ⁻1) is to the left of the y-axis and below the x-axis.

7. For numbers 7a–7b, compare. Choose <, >, or =.

7a. 0.25 ⬚ (< > =) $\frac{1}{4}$ 7b. $1\frac{1}{5}$ ⬚ (< > =) 1.5

Practice Test

6.NS.C.7a
Apply and extend previous understandings of numbers to the system of rational numbers.

1. The low weekday temperatures for a city are shown.

Low Temperatures	
Day	Low Temperature (°F)
Monday	⁻5
Tuesday	⁻3
Wednesday	2
Thursday	⁻7
Friday	3

Part A

Using the information in the table, order the temperatures from lowest to highest.

_____

Part B

Explain how to use a vertical number line to determine the order.

2. Choose <, >, or =.

2a. 0.25 | < > = | 3/4   2c. 2 7/8 | < > = | 2.875

2b. 1/3 | < > = | 0.325   2d. ⁻3/4 | < > = | ⁻1/2

GO ON

Name _____

3. Compare $-\frac{2}{3}$ and $-\frac{5}{9}$. Use words and numbers to explain your answer.

```




```

4. Choose $<$, $>$, or $=$.

4a. $-\frac{3}{5}$
$<$
$>$
$=$
   $-\frac{4}{5}$

4b. $-\frac{2}{5}$
$<$
$>$
$=$
   $-\frac{3}{4}$

4c. -6.5
$<$
$>$
$=$
   -4.2

4d. -2.4
$<$
$>$
$=$
   -3.7

5. Compare $-\frac{1}{5}$ and -0.9. Which number is greater? Use numbers and words to explain your answer.

```




```

1. Golf scores compared to par are shown.

Golf Scores	
Player	Score
Alex	⁻4
Bart	⁻1
Cal	3
Deon	⁻2

Part A

Using the information in the table, order the scores from lowest to highest.

_____

Part B

Explain how to use a horizontal number line to determine the order.

2. Four friends played a new game and Vance kept score.

Player	Score
Lou	25
Mary	⁻20
Nina	⁻30
Otto	15

When the game was finished, Vance wrote the scores in order from lowest to highest. Is Vance correct? Use words and numbers to explain why or why not. If Vance is incorrect, what is the correct order?

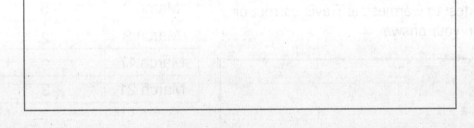

| ⁻30, ⁻20, 15, 25 |

GO ON

3. Choose <, >, or =.

3a.   1.75 meters [< / > / =] $1\frac{3}{4}$ meters   3c. $3\frac{7}{8}$ k [< / > / =] 3.375 k

3b. $\frac{^{-}2}{3}$ point [< / > / =] $^{-}0.667$ point   3d. $\frac{^{-}3}{8}$ ft [< / > / =] $\frac{^{-}1}{2}$ ft

4. Jasmine recorded the low temperatures for 3 cities.

City	Temperature (°F)
A	6
B	$^{-}4$
C	2

Draw a dot on the number line to represent the low temperature of each city. Write the letter of the city above the dot.

5. Travis made a list of his town's lowest recorded temperatures in March. He wrote the temperatures in order from coldest to warmest. Is Travis correct or incorrect? Explain your answer.

Date	Temperature (°F)
March 2	5
March 9	$^{-}2$
March 17	$^{-}9$
March 21	3

$^{-}2, 3, 5, ^{-}9$

1. Jeandre said |3| equals |⁻3|. Is Jeandre correct? Draw a number line and explain your answer.

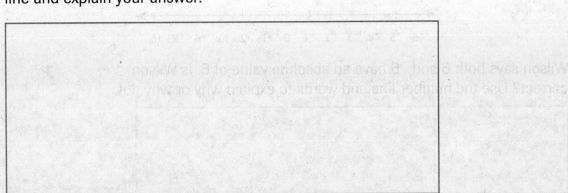

2. Tyler writes the following statements to describe the same event:

1. I added $4\frac{1}{2}$ milliliters of tank cleaner to my fishtank.

2. I had to use $4\frac{1}{2}$ milliliters of my remaining tank cleaner in order to clean my fishtank.

For Sentence 1, he graphs $4\frac{1}{2}$ on the number line.

For Sentence 2, he graphs $^-4\frac{1}{2}$ on the number line.

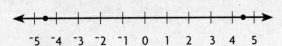

Can both points represent the event? Explain.

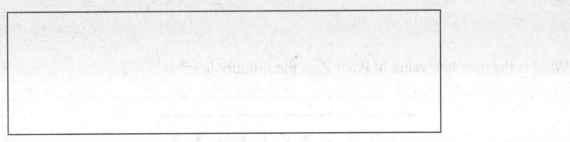

3. Graph 2 and ⁻2 on the number line.

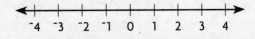

Keisha says that 2 and ⁻2 do not have the same absolute value. Is Keisha correct? Explain why or why not.

Name _____

4. Graph 6 and ⁻6 on the number line.

Wilson says both 6 and ⁻6 have an absolute value of 6. Is Wilson correct? Use the number line and words to explain why or why not.

---

5. Which point on the number line has an absolute value of 3? Mark all that apply.

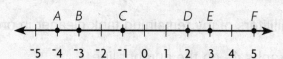

Ⓐ *A*

Ⓑ *B*

Ⓒ *D*

Ⓓ *E*

6. What is the absolute value of Point *Z* on the number line?

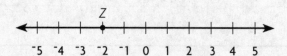

Ⓐ ⁻2

Ⓑ 0

Ⓒ 1

Ⓓ 2

Practice Test

6.NS.C.7d
Apply and extend previous understandings of numbers to the system of rational numbers.

1. Write the values in order from least to greatest.

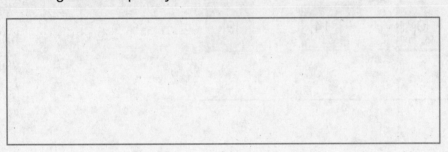

_____   _____   _____   _____

2. Lindsay and Will have online accounts for buying music. Lindsay's account balance is ⁻$20 and Will's account balance is ⁻$15. Express each account balance as a debt. Tell whose debt is greater. Explain your answer.

_____

3. Write the values in order from least to greatest.

_____

4. Write the values in order from least to greatest.

         |⁻7|

_____

GO ON

Name _____

5. Roger and Mary have an online account for buying movies. Roger's account balance is ⁻$25 and Mary's is ⁻$10. Express each account balance as a debt. Tell whose debt is greater. Explain your answer.

6. Write the values in order from least to greatest.

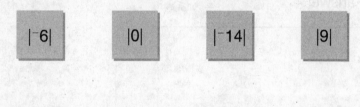

| ⁻6 | | 0 | | ⁻14 | | 9 |

_____   _____   _____   _____

7. Devon and Gwen have an account at an online bookstore. Devon's account balance is ⁻$15 and Gwen's account balance is ⁻$5. Whose account balance shows the greater debt? Explain your answer.

8. Marvin has an online account for playing games. In March, his account balance was ⁻$6. In May, his account balance was ⁻$3. Express the account balances as a debt and tell in which month Marvin's debt was greater.

1. The map shows the location *J* of Jose's house and the location *F* of the football field. Jose is going to go to Tyrell's house and then the two of them are going to go to the football field for practice.

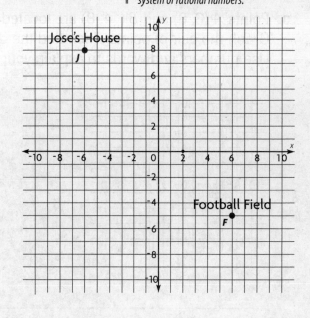

Part A

Tyrell's house is located at point *T*, the reflection of point *J* across the *y*-axis. What are the coordinates of points *T*, *J*, and *F*?

_____

Part B

If each unit on the map represents 1 block, what was the distance Tyrell traveled to the football field and what was the distance Jose traveled to the football field? Use numbers and words to explain your answer.

2. Select the pairs of points that have a distance of 10 units between them. Mark all that apply.

Ⓐ (3, ⁻6) and (3, 4)

Ⓑ (⁻3, 8) and (7, 8)

Ⓒ (4, 5) and (6, 5)

Ⓓ (4, 1) and (4, 11)

Name _____

3. Points *A* (3, 8) and *B* (⁻4, 8) are located on a coordinate plane. Graph the pair of points. Then find the distance between them. Use numbers and words to explain your answer.

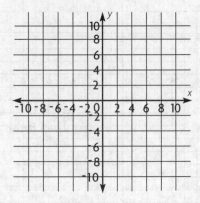

4. A map of the city hosting the Olympics is placed on a coordinate plane. Olympic Stadium is located at the origin of the map. Each unit on the map represents 2 miles.

Graph the locations of the four Olympic sites listed in the table.

Max said the distance between the Aquatics Center and the Olympic Village is greater than the distance between the Media Center and the Basketball Arena. Do you agree with Max? Use words and numbers to support your answer.

Building	Location
Olympic Village	(⁻8, 4)
Aquatics Center	(8, 4)
Media Center	(4, ⁻5)
Basketball Arena	(⁻8, ⁻5)

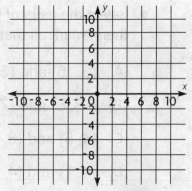

_____

_____

_____

_____

1. Ms. Hall wrote the expression $2 \times (3 + 5)^2 \div 4$ on the board. Shyann said the first step in evaluating the expression is to evaluate 5^2. Explain Shyann's mistake. Then evaluate the expression.

_____

_____

2. Select the expressions that are equivalent to $\frac{8}{27}$. Mark all that apply.

Ⓐ $\left(\frac{2}{3}\right)^3$

Ⓑ $\left(\frac{1}{27}\right)^8$

Ⓒ $\left(\frac{4}{3}\right)^2$

Ⓓ $\left(\frac{2}{3}\right)^2 \times \frac{2}{3}$

3. Use exponents to write the expression.

$3 \times 3 \times 3 \times 3 \times 5 \times 5$

$3^{\boxed{}} \times 5^{\boxed{}}$

4. Write 4.2^3 using repeated multiplication. Then find the value of 4.2^3.

_____

5. Mr. Ruiz writes the expression $5 \times (2 + 1)^2 \div 3$ on the board. Chelsea says the first step is to evaluate 1^2. Explain Chelsea's mistake. Then evaluate the expression.

GO ON

Name _____

6. Cari evaluates the expression $(5 + 4)^2 - 5 \times 2$.

Part A

Cari shows her work on the board. Use numbers and words to explain her mistake.

$(5 + 4)^2 - 5 \times 2$

$(5 + 16) - 5 \times 2$

$21 - 5 \times 2$

16×2

32

Part B

Evaluate the expression $(5 + 4)^2 - 5 \times 2$ using the order of operations. Show your work.

1. Which expressions can be used to represent "two-thirds of a number multiplied by one half"? Mark all that apply.

Ⓐ $\frac{1}{2}z \div \frac{2}{3}$

Ⓑ $\frac{2z}{3} \times \frac{1}{2}$

Ⓒ $\frac{2}{3} \times \frac{1}{2}z$

Ⓓ $\frac{2}{3z} \times \frac{1}{2}$

2. Jake writes this word expression.

the product of 7 and *m*

Write an algebraic expression for the word expression. Then, evaluate the expression for $m = 4$. Show your work.

3. An online sporting goods store charges $12 for a pair of athletic socks. Shipping is $2 per order.

Part A

Write an expression that Hana can use to find the total cost in dollars for ordering *n* pairs of socks.

_____

Part B

Hana orders 3 pairs of athletic socks, and her friend Charlie orders 2 pairs of athletic socks. What is the total cost, including shipping, for both orders? Show your work.

4. An online store sells specialty bags. They charge $8 for shipping and $21 per bag ordered. Write an expression that can be used to find the cost in dollars for *b* bags including shipping.

_____

5. Which expression can be used to represent "three times more than the sum of a number and 12.75"?

Ⓐ 3(k + 12.75)

Ⓑ 3k + 12.75

Ⓒ k + 12.75 × 3

Ⓓ 3k × (k + 12.75)

6. Sam is 5 centimeters taller than Olivia. Select the expressions that represent Sam's height if Olivia's height is *h* centimeters. Mark all that apply.

Ⓐ *h* + 5   Ⓒ *h* increased by 5

Ⓑ *h* − 5   Ⓓ *h* less than 5

7. Which expresses the calculation *add 3.7 to b*?

Ⓐ *b* + 3.7

Ⓑ *b* − 3.7

Ⓒ 3.7*b*

Ⓓ *b* ÷ 3.7

Practice Test

6.EE.A.2b
*Apply and extend previous
understandings of arithmetic to algebraic
expressions.*

 1. Kennedy bought *a* pounds of almonds at $5 per pound and
p pounds of peanuts at $2 per pound. Write an algebraic
expression for the cost of Kennedy's purchase.

_____

 2. Jasmine is buying beans. She bought *r* pounds of red beans
that cost $3 per pound and *b* pounds of black beans that cost
$2 per pound. The total amount of her purchase is given by the
expression $3r + 2b$. Select the terms of the expression. Mark all
that apply.

(A) 2

(B) 2*b*

(C) 3

(D) 3*r*

 3. Darryl is buying apples and bananas. He bought *a* pounds of
apples that cost $2 per pound and *b* pounds of bananas that
cost $1 per pound. The total amount of her purchase is given by
the expression $2a + b$. Select the terms of the expression. Mark
all that apply.

(A) 2*a*

(B) 2

(C) *a*

(D) *b*

(E) 1*b*

GO ON

4. Circle the terms in the expression. Explain how you know they are terms.

$$\frac{2}{5}a + a^2 - 9 \div 2$$

_____

_____

_____

 5. Circle the terms in the expression. Then explain how you know your answer is correct.

$6.44 \div 2 - a^2$

 6. Jasmine bought some pounds of apples at $3.75 per pound and some pounds of bananas at $1.99 per pound. If she bought the same number of pounds for each, write an algebraic expression for the cost of Jasmine's purchase.

_____

7. Elliot bought some grapes. He bought *x* pounds of red grapes that cost $4 per pound and *y* pounds of green grapes that cost $2 per pound. He used the expression $4x + 2y$ to describe the total amount of his purchase. What are the terms in the expression? Mark all that apply.

Ⓐ 4

Ⓑ $4x$

Ⓒ 2

Ⓓ $2y$

Name _____

1. The surface area of a cube can be found by using the formula $6s^2$, where s represents the length of the side of the cube.

The surface area of a cube that has a side length of

3 meters is
$$\begin{array}{|c|} \hline 54 \\ 108 \\ 2{,}916 \\ \hline \end{array}$$ meters squared.

2. Choose the number that makes the sentence true.

The formula $V = s^3$ gives the volume V of a cube with side length s. The volume of a cube that has a side length of 3.3 inches

is
$$\begin{array}{|c|} \hline 9.9 \\ 10.89 \\ 35.937 \\ \hline \end{array}$$ inches cubed.

3. What is the value of the expression $88c + 12c \times 8$ when $c = \frac{1}{2}$?

_____

4. Logan works at a florist. He earns \$600 per week plus \$5 for each floral arrangement he delivers. The expression $600 + 5f$ gives the amount in dollars that Logan earns for delivering f floral arrangements. How much will Logan earn if he delivers 45 floral arrangements in one week? Show your work.

Name _____

5. A bike rental company charges $10 to rent a bike plus $2 for each hour the bike is rented. An expression for the total cost of renting a bike for h hours is $10 + 2h$. Complete the table to find the total cost of renting a bike for h hours.

Number of Hours, h	$10 + 2h$	Total Cost
1	$10 + 2 \times 1$	
2		
3		
4		

6. Olivia delivers packages. She earns $300 per week plus $6.25 for each package she delivers. The expression $300 + 6.25p$ gives the amount in dollars that Olivia earns for delivering p packages. How much will Olivia earn if she delivers 55 packages in one week? Show your work.

7. Simon wrote the expression $x - 2 = 7$.

Which step would Simon need to take to find the value of the variable x?

(A) $x - 2 = 7 + 2$

(B) $x - 2 + 2 = 7 + 2$

(C) $x + 7 = {}^-2$

(D) $x - 2 - 7 = 0$

1. Vincent is ordering accessories for his surfboard. A set of fins costs $24 each, and a leash costs $15. The shipping cost is $4 per order. The expression $24b + 15b + 4$ can be used to find the cost in dollars of buying b fins and b leashes plus the cost of shipping.

   Select the expressions that are equivalent to $24b + 15b + 4$. Mark all that apply.

   Ⓐ $24b - 15b - 4$　　　　Ⓓ $3(8b + 5b) + 4$

   Ⓑ $24b + 19b$　　　　　　Ⓔ $(24b + 4) + (15b + 4)$

   Ⓒ $39b + 4$

2. Write the algebraic expression in the box that shows an equivalent expression.

   | $6(z + 5)$ | $6z + 5z$ | $2 + 6z + 3$ |

$6z + 5$	$11z$	$6z + 30$

3. Sora has some bags that each contain 12 potatoes. She takes 3 potatoes from each bag. The expression $12p - 3p$ represents the number of potatoes p left in the bags. Simplify the expression by combining like terms. Draw a line to match the expression with the simplified expression.

   • 　 $15p$

   • 　 $13p$

   $12p - 3p$ 　•

   • 　 $11p$

   • 　 $9p$

GO ON

4. Write the algebraic expression in the box that shows an equivalent expression.

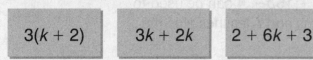

3(k + 2)	3k + 2k	2 + 6k + 3

6k + 5	5k	3k + 6

5. Emir is ordering sets of guitar strings and bags of picks for his guitar. A new set of strings costs $12, and a new bag of picks cost $4. Shipping costs $6. The expression $12g + 4g + 6$ gives the total cost for buying g sets of strings and picks. Simplify the expression by combining like terms.

_____

6. Draw a line to match the property with the equivalent expression.

Associative Property of Addition •   • $11 + (1 + c) = (11 + 1) + c$

Commutative Property of Addition •   • $0 + 11 = 11$

Identity Property of Addition •   • $11 + c = c + 11$

7. Choose the word that makes the sentence true.

Paulo wrote the expression $3 \times (d + 5)$ in his notebook. He uses the

Commutative
Associative
Distributive

Property to write the equivalent expression $3d + 15$.

1. Select the expressions that are equivalent to $3(x + 2)$. Mark all that apply.

   Ⓐ $3x + 6$

   Ⓑ $3x + 2$

   Ⓒ $5x$

   Ⓓ $x + 5$

2. Use properties of operations to determine whether $5(n + 1) + 2n$ and $7n + 1$ are equivalent expressions.

3. Alisha buys 5 boxes of peanut butter granola bars and 5 boxes of cinnamon granola bars. Let p represent the number of peanut butter granola bars and c represent the number of cinnamon granola bars. Jaira and Emma each write an expression that represents the total number of granola bars Alisha bought. Are the expressions equivalent? Justify your answer.

Jaira	Emma
$5p + 5c$	$5(p + c)$

4. Use properties of operations to determine whether $4(n + 2) + 2n$ and $6n + 2$ are equivalent expressions.

GO ON

Name _____

5. Myles bought 4 rose bushes and 4 tulip plants. Let r represent the number of roses in bloom on each rose bush and t represent the number of tulips in bloom on each tulip plant. Myles and Jenna each wrote an expression that represents the total number of flowers in bloom. Are the expressions equivalent? Justify your answer.

Myles	Jenna
$4r + 4t$	$4(r + t)$

6. Identify each expression as either Represents or Does Not Represent the surface area of the cube.

$s^3 \quad 6s \quad 4s^4 \quad 6s^2 \quad 2(s^2) + 4(s^2) \quad s^2 + s^2 + s^2 + s^2 + s^2 + s^2$

Represents	Does Not Represent

7. Use the Distributive Property to write two equivalent expressions that represent the area of the diagram. Choose from the numbers and symbols below the diagram. Not all choices will be used.

$(x + 3) \quad 2 \quad - \quad 2x \quad 6x \quad + \quad 6 \quad 3 \quad =$

7a. Represent the area as the sum of exactly two terms.

7b. Represent the area as a product in which one factor is a sum.

_____   _____

7c. Explain why the expressions are equivalent.

Name _____

1. In the inequality $\frac{2}{5}v \geq 10$, determine whether $v \geq 25$ is a solution. Explain your answer.

2. The distance from third base to home plate is 88.9 feet. Romeo was 22.1 feet away from third base when he was tagged out. The equation $88.9 - t = 22.1$ can be used to determine how far he needed to run to get to home plate. Using substitution, the coach determines that Romeo needed

to run
| 66 |
| 66.8 |
| 111 |
feet to get to home plate.

3. The maximum number of players allowed on a lacrosse team is 23. The inequality $t \leq 23$ represents the total number of players, t, allowed on the team.

Two possible solutions for the inequality are
| 23 |
| 25 |
| 27 |
and
| 26. |
| 24. |
| 22. |

4. Mr. Charles needs to have at least 10 students sign up for homework help in order to use the computer lab. The inequality $h \geq 10$ represents the number of students, h, who must sign up. Select possible solutions of the inequality. Mark all that apply.

Ⓐ 7   Ⓓ 10

Ⓑ 8   Ⓔ 11

Ⓒ 9   Ⓕ 12

Name _____

5. The marking period is 45 school days long. Today is the twenty-first day of the marking period. The equation $x + 21 = 45$ can be used to find the number of days left in the marking period. Using substitution, Rachel determines

there are
20
24
26
days left in the marking period.

6. In a basket of fruit, $\frac{5}{6}$ of the pieces of fruit are apples. There are 20 apples in the display. The equation $\frac{5}{6}f = 20$ can be used to find how many pieces of fruit f are in the basket. Use words and numbers to explain how to solve the equation to find how many pieces of fruit are in the basket.

_____

_____

_____

7. Use exponents to write the expression.

$2 \times 2 \times 2 \times 2 \times 2 \times 4 \times 4 \times 4$

$2^{\square} \times 4^{\square}$

8. Write the algebraic expression in the box that shows an equivalent expression.

$2(j + 3)$	$2j + 3j$	$3 + 6j + 2$

$5j$	$5 + 6j$	$2j + 6$

1. A plumber charges $10 for transportation and $55 per hour for repairs. Write an expression that can be used to find the cost in dollars for a repair that takes *h* hours.

_____

2. Ellen is 2 years older than her brother Luke. Let *k* represent Luke's age. Identify the expression that can be used to find Ellen's age.

   Ⓐ *k* − 2   Ⓒ 2*k*

   Ⓑ *k* + 2   Ⓓ $\frac{k}{2}$

3. Abe is 3 inches taller than Chen. Select the expressions that represent Abe's height if Chen's height is *h* inches. Mark all that apply.

   Ⓐ *h* − 3   Ⓒ the sum of *h* and 3

   Ⓑ *h* + 3   Ⓓ the difference between *h* and 3

4. Erika writes the word expression

   the quotient of 24 and *k* .

   Write an algebraic expression for the word expression. Then, evaluate the expression for *k* = 3. Show your work.

GO ON

Name _____

5. A resort rents surfboards for $15 plus $3 for each hour the surfboard is rented. An expression for the total cost of renting a surfboard for h hours is $15 + 3h$. Complete the table by finding the total cost of renting a surfboard for h hours.

Number of hours, h	$15 + 3h$	Total Cost
1		
2		
3		
4		

6. An online camping supplies outlet charges $14 for a canteen, and shipping is $2 per order.

Part A

Write an expression that CJ can use to find the total cost in dollars for ordering n canteens.

_____

Part B

CJ orders 2 canteens, and his friend Cameron orders 4 canteens. What is the total cost, including shipping, for both orders? Show your work.

1. Match each scenario with the equation that can be used to solve it.

> Jane's dog eats 3 pounds of food a week. How many weeks will a 24-pound bag last?

• 3x = 39

> There are 39 students in the gym, and there are an equal number of students in each class. If 3 classes are in the gym, how many students are in each class?

• 4x = 24

> There are 4 games at the carnival. Kevin played all the games in 24 minutes. How many minutes did he spend at each game if he spent an equal amount of time at each?

• 3x = 24

2. Bryan rides the bus to and from work on the days he works at the library. In one month, he rode the bus 24 times. Solve the equation 2x = 24 to find the number of days Bryan worked at the library. Draw a model.

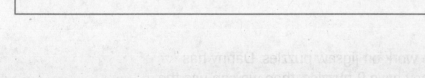

Name _____

3. Malorie uses $\frac{2}{3}$ foot of string to make a bracelet. She bought 6 feet of string.

Part A

Write and solve an equation to find how many bracelets x she can make from 6 feet of string.

Part B

Explain how you determined which operation was needed to write the equation.

4. Suzan's tulips are 6.4 inches shorter than her rose bush. The rose bush is 13.2 inches tall. Write and solve an addition equation to find the height of her tulips.

5. Danny and Carly like to work on jigsaw puzzles. Danny has 2 puzzles. If together they have 9 puzzles, then we can use the equation $x + 2 = 9$ to determine how many puzzles Carly has. How many puzzles does Carly have?

1. The maximum capacity of the school auditorium is 420 people. Write an inequality for the situation. Tell what type of numbers the variable in the inequality can represent.

_____

2. Match the inequality to the word sentence it represents.

$w < 70$ •

• The temperature did not drop below 70 degrees.

$x \leq 70$ •

• Dane saved more than \$70.

$y > 70$ •

• Fewer than 70 people attended the game.

$z \geq 70$ •

• No more than 70 people can participate.

3. Cydney graphed the inequality $d \leq -14\frac{1}{2}$.

−14

Part A

Dylan said that $-14\frac{1}{2}$ is not part of the solution of the inequality. Do you agree or disagree with Dylan? Use numbers and words to support your answer.

Part B

Suppose Cydney's graph had an empty circle at $-14\frac{1}{2}$. Write the inequality represented by this graph.

_____

GO ON

Name _____

4. The minimum wind speed for a storm to be considered a hurricane is 74 miles per hour. The inequality $w \geq 74$ represents the possible wind speeds of a hurricane.

Two possible solutions for the inequality $w \geq 74$

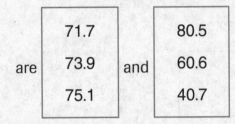

are
| 71.7 |
| 73.9 |
| 75.1 |

and
| 80.5 |
| 60.6 |
| 40.7 |

5. Match the inequality with the word sentence it represents.

$r > 10$ •   • | Walter sold more than 10 tickets. |

$s \leq 10$ •   • | Fewer than 10 children are at the party. |

$t \geq 10$ •   • | No more than 10 people can be seated at a table. |

$w < 10$ •   • | At least 10 people need to sign up for the class. |

6. Alena graphed the inequality $c \leq 25$.

Darius said that 25 is not part of the solution of the inequality. Do you agree or disagree with Darius? Use numbers and words to support your answer.

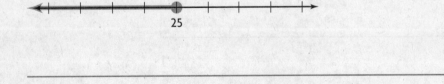

25

_____

_____

1. A box of peanut butter crackers contains 12 individual snacks. The total number of individual snacks *s* is equal to 12 times the number of boxes of crackers *b*.

The independent variable is

> *b.*
> *s.*

The dependent variable is

> *b.*
> *s.*

The equation that represents the relationship between the variables is

> $b = 12s.$
> $s = 12b.$

2. A stationery store charges $8 to print logos on paper purchases. The total cost *c* is the price of the paper *p* plus $8 for printing the logo.

Choose the word that makes the statement correct.

p is the

> linear
> dependent
> independent

variable.

3. Miranda's wages are $15 per hour. Write a linear equation that gives the wages *w* in dollars that Miranda earns in *h* hours.

_____

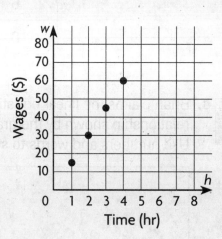

GO ON

4. Alex swims 20 minutes per day for exercise. The total number of minutes *m* she swims equals 20 times the number of days *d* she swims.

What is the dependent variable?

What is the independent variable?

Write the equation that represents the relationship between the *m* and *d*.

5. To rent a beach chair and an umbrella, there is a rental fee of $10. Then, it costs $2 per day. Use the equation $c = 2d + 10$ to complete the table.

Input	Output
Days, *d*	Cost ($), *c*
2	
4	
6	
8	

6. Brian claims the linear equation for the relationship shown by the graph is $c = 35d$. Use numbers and words to support Brian's claim.

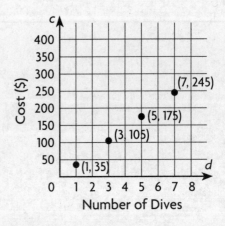

Name _____

1. Find the area of the parallelogram.

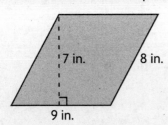

7 in.   8 in.

9 in.

The area is _____ in.²

2. A wall tile is two different colors. What is the area of the white part of the tile? Explain how you found your answer.

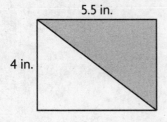

5.5 in.

4 in.

3. A carpenter needs to replace some flooring in a house.

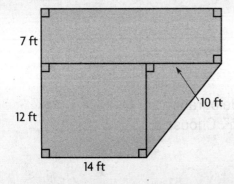

7 ft

12 ft

10 ft

14 ft

Select the expression that can be used to find the total area of the flooring to be replaced. Mark all that apply.

Ⓐ 19 × 14

Ⓒ 19 × 24 − $\frac{1}{2}$ × 10 × 12

Ⓑ 168 + 12 × 14 + 60

Ⓓ 7 × 24 + 12 × 14 + $\frac{1}{2}$ × 10 × 12

GO ON

Name _____

 4. A trapezoid has an area of 30 in.² If the lengths of the bases are 4.8 in. and 5.2 in., what is the height?

_____ in.

 5. A quilt is in the shape of a regular pentagon. It is made from 5 pieces of fabric that are congruent triangles. Each triangle has an area of 16 in.² What is the area of the quilt?

_____ in.²

 6. Name the polygon and find its area. Show your work.

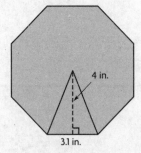

4 in.

3.1 in.

polygon: _____   area: _____

 7. The roof of Braeden's house is shaped like a parallelogram. The base of the roof is 12 m and the area is 114 m². Choose a number and unit to make a true statement.

The height of the roof is

102		m
30		m²
19		m³
9.5		

.

1. A prism is filled with 44 cubes with $\frac{1}{2}$-unit side lengths. What is the volume of the prism in cubic units?

_____ cubic units

2. Dominique has a box of sewing buttons that is in the shape of a rectangular prism.

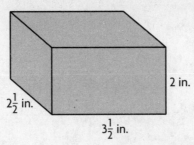

2 in.

$2\frac{1}{2}$ in.

$3\frac{1}{2}$ in.

The volume of the box is $2\frac{1}{2}$ in. × $3\frac{1}{2}$ in. ×

2 in.	8 in.³
$2\frac{1}{2}$ in.	$17\frac{1}{2}$ in.³
$3\frac{1}{2}$ in.	35 in.³

3. Select the following expressions that can be used to find the volume of the rectangular prism. Mark all that apply.

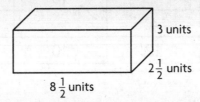

3 units

$2\frac{1}{2}$ units

$8\frac{1}{2}$ units

(A) $2\frac{1}{2}$ units × $8\frac{1}{2}$ units × 3 units

(B) $4(8\frac{1}{2}$ units × 3 units$) + 2(2\frac{1}{2}$ units × 3 units$)$

(C) 63.75 cubic units

(D) $2(8\frac{1}{2}$ units × $2\frac{1}{2}$ units$) + 2(8\frac{1}{2}$ units × 3 units$)$
$+ 2(2\frac{1}{2}$ units × 3 units$)$

GO ON

4. A box measures 5 units by 3 units by $2\frac{1}{2}$ units.
What is the volume of the box?

_____ cubic units

5. A box measures 4 units by $2\frac{1}{2}$ units by $1\frac{1}{2}$ units.
What is the greatest number of cubes with a side
length of $\frac{1}{2}$ unit that can be packed inside the box?

Ⓐ 120

Ⓑ 220

Ⓒ 300

Ⓓ 1,500

6. Gary wants to build a shed shaped like a rectangular prism in his
backyard. He goes to the store and looks at several different options.
The table shows the dimensions and volumes of four different sheds.

Use the formula $V = l \times w \times h$ to complete the table.

	Length (ft)	Width (ft)	Height (ft)	Volume (ft³)
Shed 1		10	8	960
Shed 2	18		10	2,160
Shed 3	12	4		288
Shed 4	10	12	10	

7. A prism is filled with 25 cubes with $\frac{1}{2}$-unit side lengths.
What is the volume of the prism in cubic units?

☐ cubic units

1. Kareem is drawing parallelogram *ABCD* on the coordinate plane.

Find and label the coordinates of the fourth vertex *D* of the parallelogram. Draw the parallelogram.

What is the length of side *CD*? How do you know?

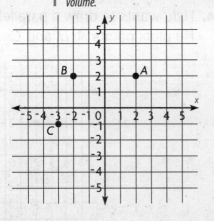

_____

_____

 2. Suppose the point (3, 2) is changed to (3, 1) on this rectangle. What other point must change so that the figure remains a rectangle? What is the area of the new rectangle?

Point : _____ would change to _____.

The area of the new rectangle is _____ square units.

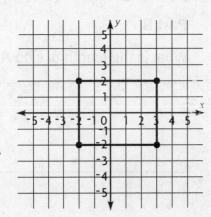

 3. Eliana is drawing a figure on the coordinate grid. Select the statements that are correct. Mark all that apply.

Ⓐ The point (⁻1, 1) would be the fourth vertex of a square.

Ⓑ The point (1, 1) would be the fourth vertex of a trapezoid.

Ⓒ The point (2, ⁻1) would be the fourth vertex of a trapezoid.

Ⓓ The point (⁻1, ⁻1) would be the fourth vertex of a square.

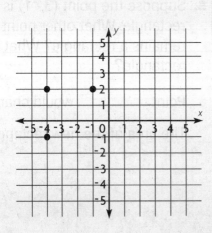

GO ON ➡

Name _____

4. Hsiu wants to draw a parallelogram on the coordinate plane. He plots points *A*, *D*, and *C*.

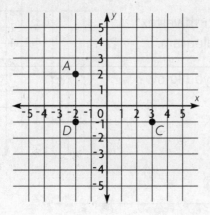

Part A

Find and label the coordinates of the fourth vertex *B* of the parallelogram. Draw the parallelogram.

Part B

What is the length of side *AB*? How do you know?

5. Suppose the point (3, ⁻1) is changed to (3, 0) on this rectangle. What other point must change so that the figure remains a rectangle? What is the area of the new rectangle?

Point _____ would change to _____.

The area of the new rectangle is _____ square units.

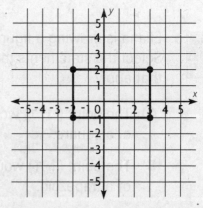

Practice Test

6.G.A.4
Solve real-world and mathematical problems involving area, surface area, and volume.

Name _____

1. Elaine makes a rectangular pyramid from paper.

The base is a
| rectangle. |
| square. |
| triangle. |

The lateral faces are
| rectangles. |
| squares. |
| triangles. |

2. Tina cut open a cube-shaped microwave box to see the net. How many square faces does this box have?

_____ square faces

3. Charles is painting a treasure box in the shape of a rectangular prism.

Which nets can be used to represent Charles's treasure box? Mark all that apply.

Ⓐ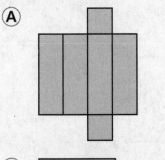

Ⓒ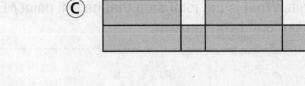

Ⓑ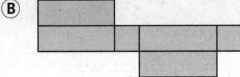

Ⓓ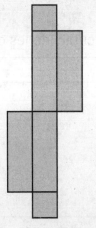

GO ON

4. Jason is covering an ottoman with fabric. The ottoman is in the shape of a rectangular prism that is 37 cm long, 21 cm wide, and 30 cm high. How much fabric is needed to cover the sides and top of the ottoman? Explain your strategy.

5. Eli made a wooden box in the shape of a rectangular prism. The box has a length of 5 inches, a width of $3\frac{1}{2}$ inches, and a height of 7 inches.

Part A

Eli wants to paint the entire box green and give it to his dad as a gift. What is the total area that he will paint? Explain how you found your answer.

Part B

Can the box hold 200 cubic inches of packing peanuts? Explain how you know.

6. A gift box measures 8 inches by 10 inches by 3 inches. What is the surface area of the box?

[   ] in.²

1. Michael's teacher asks, "How many items were sold on the first day of the fund raiser?" Explain why this is not a statistical question.

2. Select the questions that are statistical questions. Mark all that apply.

   (A) What is the height of the tallest tree in each of the national parks?

   (B) What is the difference in height between the tallest tree and the shortest tree in each of the national parks?

   (C) How tall is the cypress tree on the north side of the lake this morning?

   (D) What are the heights of the trees that are taller than 30 feet?

3. Select the questions that are statistical questions. Mark all that apply.

   (A) How many minutes did it take Ethan to complete his homework last night?

   (B) How many minutes did it take Madison to complete her homework each night this week?

   (C) How many more minutes did Andrew spend on homework on Tuesday than on Thursday each week?

GO ON

4. A researcher asks, "How much electricity did Home 12 use on Day 1?" Explain why this is not a statistical question.

5. Select the questions that are statistical questions. Mark all that apply.

Ⓐ How many pets do you have in your home?

Ⓑ How tall are basketball players?

Ⓒ Who is the tallest 6th grade student?

Ⓓ How many minutes long is a lunch period in a school?

Ⓔ How much time do 6th grade students spend doing homework every night?

6. Select the questions that are statistical questions. Mark all that apply.

Ⓐ How many notebooks do you have?

Ⓑ What is the average height of an Olympic swim team?

Ⓒ What is the area of a football field?

Ⓓ How many people eat ice cream during the summer?

Ⓔ How long can students in a class hold their breath?

1. The dot plots show the number of chin-ups done by the 4 sixth grade gym classes at Sagefeld Middle School.

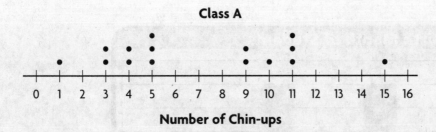

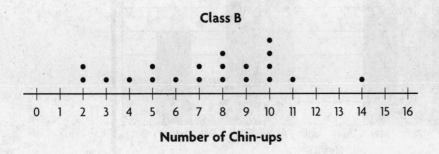

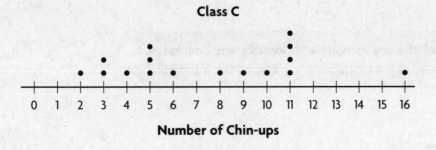

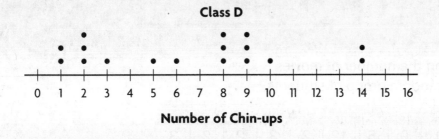

For which 2 classes do the dot plots show data that have the most similar means, medians, and ranges?

_____

GO ON

Name _____

2. Mrs. Gutierrez made a histogram of the birth month of the students in her class. Describe the patterns in the histogram by completing the chart.

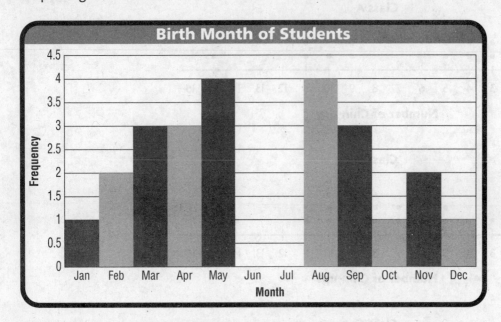

Birth Month of Students

Identify any peaks.	Identify any increases across the intervals.	Identify any decreases across the intervals.

3. Diego collected data on the number of movies seen last month by a random group of students.

Number of Movies Seen Last Month												
0	1	3	2	1	0	5	12	2	3	2	2	3

Draw a box plot of the data and use it to find the interquartile range and range.

Interquartile range _____

Range _____

0 1 2 3 4 5 6 7 8 9 10 11 12 13 14

Number of Movies Seen Last Month

1. Kylie's teacher collected data on the heights of boys and girls in a sixth grade class. Use the information in the table to compare the data.

Heights (in.)							
Girls	55	60	56	51	60	63	65
Boys	72	68	70	56	58	62	64

The mean of the boys' heights is ⸢ the same as / less than / greater than ⸣ the mean of the girls' heights.

The range of the boys' heights is ⸢ the same as / less than / greater than ⸣ the range of the girls' heights.

2. The box plot shows the number of boxes of paper sold at an office supply store each day for a week.

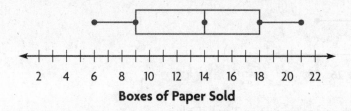

Boxes of Paper Sold

For numbers 2a–2c write whether the statement is True, False, or if there is Not Enough Information.

2a. The median is 18. _____

2b. The range is 15. _____

2c. The interquartile range is 9. _____

3. Jake's final grade in his science class is calculated by finding the mean of his scores for six project reports. The scores Jake received on his first five reports are 66, 80, 88, 82, and 72.

What is the lowest possible score Jake can earn on his last report in order to have at least an 80 for his final grade?

(A) 86

(C) 92

(B) 88

(D) 96

GO ON

Name _____

4. Calculate the range and interquartile range for the data displayed in the dot plot. Show your work.

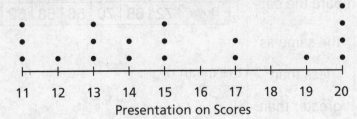

Presentation on Scores

Range: _____

Interquartile range: _____

_____

5. The box plot shows the number of points scored in each game by a football team one season.

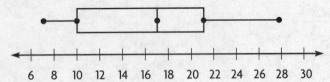

For numbers 5a–5d write whether the statement is True, False, or if there is Not Enough Information.

5a.   The range is 22.   _____

5b.   The mean is 18.   _____

5c.   The lower quartile is 10.   _____

5d.   The interquartile range is 11.   _____

STOP

Name _____

1. The data set shows the ages of the members of the cheerleading squad. Plot the data on the dot plot. What is the most common age of the members of the squad? Explain how you found your answer.

Ages of Cheerleaders (years)				
8	11	13	12	14
12	10	11	9	11

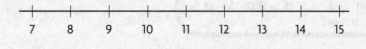

2. Ian collected data on the number of children in 13 different families.

Number of Children												
1	2	4	3	2	1	0	8	1	1	0	2	3

Draw a box plot of the data and use it to find the interquartile range and range.

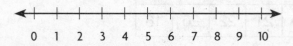

Interquartile range: _____   Range: _____

GO ON

3. The frequency table shows the TV ratings for the show *American Singer*. Complete the histogram for the data.

TV Ratings	
Rating	Frequency
14.1–14.5	2
14.6–15.0	6
15.1–15.5	6
15.6–16.0	5
16.1–16.5	1

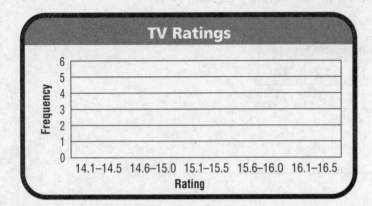

4. The data set shows the total points scored by the middle school basketball team in the last 14 games. What is the most common number of points scored in a game? Explain how to find the answer using a dot plot.

Total Points Scored						
42	36	35	49	52	43	41
32	45	39	50	38	37	39

_____

_____

5. The data set shows the number of desks in 12 different classrooms.

Classroom Desks											
24	21	18	17	21	19	17	20	21	22	20	16

Find the values of the points on the box plot.

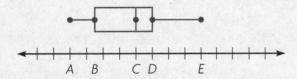

$A =$ ☐ $B =$ ☐ $C =$ ☐ $D =$ ☐ $E =$ ☐

1. Describe the data set by writing the attribute measured, the unit of measure, the likely means of measurement, and the number of observations in the correct location on the chart.

Heights of Sixth Graders (in.)						
50	58	56	60	58	52	50
53	54	61	48	59	48	59
55	59	62	49	57	56	61

21

yardstick

inches

heights of sixth graders

Attribute	Unit of Measure	Likely Means of Measurement	Number of Observations

2. Describe the data set by writing the attribute measured, the unit of measure, the likely means of measurement, and the number of observations in the correct location on the chart.

Daily Temperature (°F)						
64	53	61	39	36	43	48

7   thermometer   degrees Fahrenheit   daily temperature

Attribute	Unit of Measure	Likely Means of Measurement	Number of Observations

GO ON

3. A teacher surveys her students to find out how much time the students spent eating lunch on Monday.

She uses | hours |
| minutes | as the unit of measure.
| seconds |

Monday Lunch Time (min.)			
15	18	18	14
15	20	16	15
15	19	15	19

 4. Describe the data set by writing the attribute measured, the unit of measure, the likely means of measurement, and the number of observations in the correct location on the chart.

100-Meter Run Data						
12.8 seconds	12.5 seconds	12.9 seconds	13.4 seconds	13.5 seconds	13.7 seconds	12.8 seconds

7	stopwatch	seconds	time to run a 100-meter race

Attribute	Unit of Measure	Likely Means of Measurement	Number of Observations

 5. A teacher surveys her students to find out how much time the students spent completing their art projects.

She uses | hours |
| minutes | as the unit of measure.
| seconds |

Art Project Time (min.)			
35	50	25	30
20	15	55	30
20	35	50	15

 6. Debra surveys her classmates to find out how much time each night they spend sleeping.

Time Spent Sleeping (hr.)			
7	9	9	6
8	7	8	6
7	9	9	8

She uses | hours |
| minutes | as the unit of measure.
| seconds |

1. The numbers of sit-ups students completed in one minute are 10, 42, 46, 50, 43, and 49. The mean of the data values is 40 and the median is 44.5. Which measure of center is less affected by the outlier of 10, the mean or median? Use words and numbers to support your answer.

_____

_____

_____

2. The Martin family goes out for frozen yogurt to celebrate the last day of school. The costs of their frozen yogurts are $1, $1, $2, and $4. Choose the word that makes the statement correct.

The | mean
     | median | cost for the frozen yogurts can be found by adding
     | mode

each cost and dividing that total by 4.

3. Larry is training for a bicycle race. He records how far he rides each day in a table. Find the mode of the data.

Miles Larry Rides Each Day					
Monday	Tuesday	Wednesday	Thursday	Friday	Saturday
15	14	12	16	15	15

_____

GO ON

4. The dot plot shows the number of errors made by a baseball team in the first 16 games of the season. Are there any modes in the data? How many? Explain your answer.

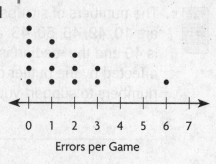

Errors per Game

5. The data set shows the number of soccer goals scored by players in 3 games. Which statements about the data are correct? Mark all that apply.

Number of Goals Scored			
Player A	1	2	1
Player B	2	2	2
Player C	3	2	1

(A) The mean absolute deviation of Player A is 1.

(B) The mean absolute deviation of Player B is 0.

(C) The mean absolute deviation of Player C is greater than the mean absolute deviation of Player A.

6. The box plot shows the heights of corn stalks from two different farms.

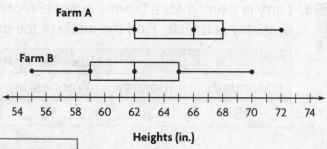

Heights (in.)

The range of Farm A's heights is

| the same as |
| less than |
| greater than |

the range of Farm B's heights.

1. The numbers of points scored by a football team in 7 different games are 26, 38, 33, 20, 27, 3, and 28. Which shows the difference between the outlier and the highest score in the data set?

    (A) 17

    (B) 25

    (C) 28

    (D) 35

2. The amounts of money Connor earned each week from mowing lawns for 5 weeks are $12, $61, $71, $52, and $64. The mean amount earned is $52 and the median amount earned is $61. Identify the outlier and describe how the mean and median for this set of data are affected by the outlier.

3. The prices of mesh athletic shorts at five different stores are $9, $16, $18, $20, and $22. The mean price is $17 and the median price is $18. Identify the outlier and describe how the mean and median for this set of data are affected by it.

4. The numbers of miles Madelyn drove between stops were 182, 180, 181, 184, 198, and 185. Which measure of center is a value that is less than 3 of the data values?

    (A) mean   (C) mode

    (B) median   (D) range

GO ON

Name _____

5. The amounts of money Brittany earned each week from babysitting for 5 weeks are $12, $62, $70, $54, and $62. The mean amount earned is $52 and the median amount earned is $62. Identify the outlier and describe how the mean and median for this data set are affected by the outlier.

6. The high temperatures for the week in Cincinnati, in degrees Fahrenheit, were 43, 43, 45, 42, 26, 43, and 45. Choose the word that makes the statement correct.

The outlier in this set of data affects the mean

by | decreasing / equalling / increasing | it.

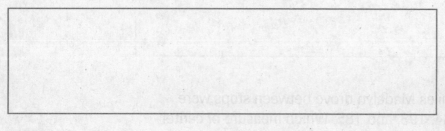

7. The numbers of emails Reese received each hour are 8, 7, 10, 8, 1, 9, 8, and 11. The mean of the data values is 7.75 and the median is 8.5. If any one of the data values were higher or lower, the mean would change. Is there any single value that could change which would affect the median? Use words and numbers to support your answer.

8. Luis' bowling scores were 195, 194, 191, 190, 208, 190, and 192. Which measure of center is equal to the least value in the data set?

(A) mean   (C) mode

(B) median   (D) range

1. Divide.

$$\frac{7}{8} \div \frac{3}{5} = \boxed{}$$

2. Ashley evaluates the expression $4 \times (3 + 6)^2$ and gets 156. Is Ashley correct? Explain your answer.

_____

_____

_____

3. Determine whether each ratio is equivalent to $\frac{1}{3}$, $\frac{5}{10}$, or $\frac{3}{5}$. Write the ratio in the correct box.

$$\frac{2}{4} \qquad \frac{3}{9} \qquad \frac{7}{21} \qquad \frac{18}{30}$$

$\frac{1}{3}$	$\frac{5}{10}$	$\frac{3}{5}$

$$\frac{10}{30} \qquad \frac{6}{10} \qquad \frac{1}{2} \qquad \frac{8}{16}$$

4. The frequency table shows the ages of the students on the middle school crew team. Complete the histogram for the data.

Ages of Students on Crew Team	
9–10	2
11–12	6
13–14	8
15–16	4

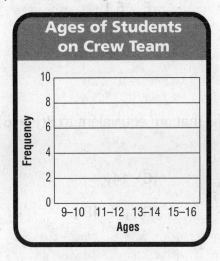

GO ON ➡

5. Kelly collected $15, $15, $25, and $29 for the class fundraiser.

The mean of the donations is
| $15. |
| $21. |
| $25. |
The median of the donations

is
| $20. |
| $21. |
| $25. |
The mode of the donations is
| $15. |
| $20. |
| no mode. |

6. At a convenience store, the Jensen family puts 12.4 gallons of gasoline in their van at a cost of $3.80 per gallon. They also spend $7.96 on water bottles and $3.10 on snacks.

Mrs. Jensen says the total cost for everything before tax is $56.66. Do you agree with her? Explain why or why not.

_____

_____

_____

7. Select the numbers that would be located between ⁻6 and ⁻5 on a number line. Mark all that apply

- (A) $6\frac{1}{4}$
- (B) $⁻5\frac{7}{8}$
- (C) ⁻6.01
- (D) 5.6
- (E) ⁻5.1

8. Select the expressions that are equivalent to $8(y + 6)$. Mark all that apply.

- (A) $8y + 48$
- (B) $8y + 6$
- (C) $14y$
- (D) $y + 48$

GO ON

9. Danica sold cookies as a fund raiser each day for 7 days. She raised $27, $25, $19, $20, $22, $23, and $11. Identify the outlier. If the outlier increased, would that affect the mean of this set of data? Explain your answer.

_____

_____

10. Identify the quadrant where each point is located. Write each point in the correct box.

(4, 6)   (7, ⁻2)   (3, 9)

(⁻5, 9)   (⁻3, ⁻6)   (⁻6, ⁻1)

Quadrant I	Quadrant II	Quadrant III	Quadrant IV

 11. Greg mixes 6 cans of black paint with 8 cans of white paint to get a gray paint. How many cans of black paint will he need to mix with 48 cans of white paint to get the same gray color?

Show your work in the space below.

_____ cans

12. Select the pairs of points that have a distance of 10 between them. Mark all that apply.

Ⓐ (2, ⁻2) and (2, 8)

Ⓑ (0, ⁻5) and (0, 5)

Ⓒ (8, 4) and (3, 4)

Ⓓ (2, ⁻4) and (2, 6)

13. Circle <, >, or =.

13a. $^-3$ $\boxed{\begin{matrix} < \\ > \\ = \end{matrix}}$ $^-5$

13c. $^-5.9$ $\boxed{\begin{matrix} < \\ > \\ = \end{matrix}}$ $^-4.5$

13b. $^-1$ $\boxed{\begin{matrix} < \\ > \\ = \end{matrix}}$ $^-7$

13d. $\frac{-1}{3}$ $\boxed{\begin{matrix} < \\ > \\ = \end{matrix}}$ $\frac{-1}{2}$

14. Match the inequality to the word sentence it represents.

$a < 60$ •

• | No more than 60 people can participate. |

$b \leq 60$ •

• | Miranda saved less than $60. |

$c > 60$ •

• | More than 60 people attended the game. |

$d \geq 60$ •

• | The temperature did not drop below 60 degrees. |

15. Identify the expression that can be used to express the calculation *add 9 to z.*

(A) $z - 9$　　　(C) $z + 9$

(B) $9z$　　　(D) $9z + 9$

16. Select the questions that are statistical questions. Mark all that apply.

(A) What are the lengths of the hiking trails in the park?

(B) How many more people hike the different trails in the fall than in the summer?

(C) How long is the trail that Steven hiked yesterday?

GO ON →

17. Olivia is sorting through the coins in her bank to determine how much money she has. She sees that she has $3.25 in quarters. The equation $0.25x = \$3.25$ can be used to figure out how many quarters are in Olivia's bank. Using substitution, Olivia

determines that she has
| 11 |
| 12 |
| 13 |
quarters.

18. The dot plot shows the number of text messages 15 students sent on a particular day. Select the statements that describe patterns in the data. Mark all that apply.

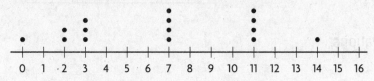

Number of Text Messages Sent

(A) There are two peaks.

(B) There are no clusters.

(C) There is a gap from 4 to 6.

(D) There is a gap from 8 to 11.

(E) The modes are 7 and 11.

19. Write the values in order from least to greatest.

|5| |⁻9| |⁻2| |7|

_____

GO ON

20. Gordon bought x boxes of granola bars at \$4 per box and y boxes of raisins at \$2 per box. Write an algebraic expression for the cost of Gordon's purchases.

_____

21. A manager at a car rental company surveys her customers to find how far the customers drive their cars over one weekend.

She uses

| yards |
| miles |
| kilometers |

as the unit of measure.

Distance Driven (miles)			
100	125	180	94
153	98	125	118
95	105	116	55

She made

| 10 |
| 11 |
| 12 |

observations.

22. Use the net to find the total surface area of the solid figure it represents. Show your work.

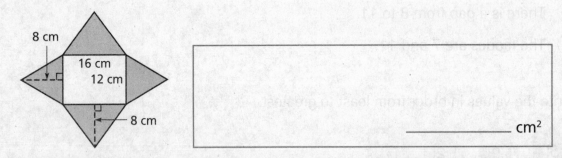

8 cm

16 cm

12 cm

8 cm

_____ cm²

23. A rectangular prism measures 5 units long, $\frac{1}{2}$ unit wide, and $5\frac{1}{2}$ units high. Which expressions use the formula for volume, $V = Bh$, to show the volume of the rectangular prism? Mark all that apply.

(A) $5\frac{1}{2}(5 \times \frac{1}{2})$

(C) 5 units $\times \frac{1}{2}$ unit $\times 5\frac{1}{2}$ units

(B) 2(5 units $\times \frac{1}{2}$ unit)

(D) 2($5\frac{1}{2}$ units $\times$ 5 units $\times \frac{1}{2}$ unit)

24. Choose the coordinate pair that makes the statement correct.

Point A
| |
| (2, ⁻1) |
| (⁻2, ⁻1) |
| (2, 1) |
| (⁻2, 1) |
 is to the right of the y-axis and below the

x-axis.

25. The area of a triangle is 30 ft². Select the pairs of dimensions that could be the height and base of the triangle. Mark all that apply.

(A) $h = 3$, $b = 10$

(B) $h = 3$, $b = 20$

(C) $h = 5$, $b = 12$

(D) $h = 5$, $b = 24$

26. For rectangle $ABCD$, use subtraction to find the length of side BC. Explain how you found the length.

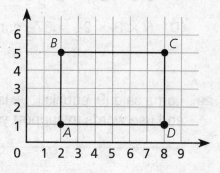

_____ units

GO ON

27. Choose one number from each column to show the mean, the median, and the range of the data set.

13, 27, 17, 7, 16, 17, 29, 10, 24, 12, 15

Mean	Median	Range
○ 12	○ 12	○ 12
○ 13	○ 13	○ 14
○ 15	○ 15	○ 15
○ 16	○ 16	○ 16
○ 17	○ 17	○ 17
○ 24	○ 24	○ 22

28. A box measures 5 units by 3 units by $2\frac{1}{2}$ units. Which shows a way to find the volume of the box?

Ⓐ $5 + 3 + 2\frac{1}{2} = 10\frac{1}{2}$ cubic units

Ⓑ $2(5 + 3 + 2\frac{1}{2}) = 21$ cubic units

Ⓒ $5(3 + 2\frac{1}{2}) = 27\frac{1}{2}$ cubic units

Ⓓ $5 \times 3 \times 2\frac{1}{2} = 37\frac{1}{2}$ cubic units

29. Lorna said |9| equals |⁻9|. Is Lorna correct? Draw a number line and use words to support your answer.

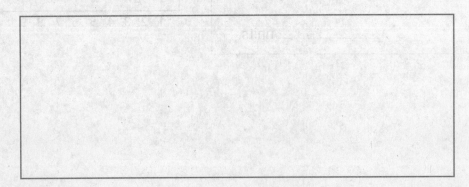

GO ON

Math Carnival

Ms. Lee's class holds a math carnival with a variety of math games and activities.

1. One of the activities is a match game. There is a set of cards with fractions on half the cards and the equivalent decimals on the other half. The cards are turned face down in an array, and players take turns picking two cards and trying to match a fraction to its equivalent decimal.

Draw a line from each fraction to its equivalent decimal. Show your work in the space below the decimal cards.

| $\frac{1}{2}$ | $\frac{1}{3}$ | $\frac{1}{4}$ | $\frac{1}{6}$ | $\frac{1}{8}$ | $\frac{2}{3}$ | $\frac{3}{4}$ | $\frac{5}{6}$ | $\frac{3}{8}$ | $\frac{5}{8}$ | $\frac{7}{8}$ | $\frac{8}{8}$ |

| 0.125 | 0.1$\overline{66}$ | 0.25 | 0.$\overline{33}$ | 0.375 | 0.5 | 0.625 | 0.$\overline{66}$ | 0.75 | 0.8$\overline{3}$ | 0.875 | 1.00 |

GO ON ➡

2. The fraction/decimal cards are also used to make sums and differences. To play the game you try to find sets of three cards that make an addition or subtraction math fact. You can use two cards with the same number to make a set.

Examples: $\frac{1}{3} + \frac{1}{6} = \frac{1}{2}$ and $\frac{1}{2} - \frac{1}{6} = \frac{1}{3}$ $0.25 + 0.25 = 0.5$ and $0.5 - 0.25 = 0.25$

Try to find as many sets as you can. Write the sets of addition and subtraction facts you find.

3. To play "Max Fractions," you have to string beads with fractions on them as fast as possible. Number these fractions in order from least to greatest. Show your work.

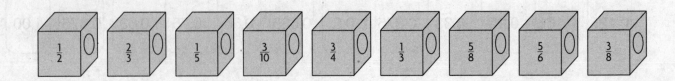

GO ON

4. Another game is to guess the number of jelly beans in a jar. The table shows the guesses made by each of the players on two teams of five. Find the average for each team to the nearest tenth.

Team A's average: _____

Team B's average: _____

The actual number of jelly beans is 271.

How far off is Team A's average? _____

How far off is Team B's average? _____

In the third column and the sixth column, write the difference between each guess and the real value. To find the difference, subtract the actual value from the guess.

guess − actual value = difference

In the fourth and seventh columns, write the absolute value of each difference.

Some answers for each team are done for you.

	Team A	Difference	Abs Val	Team B	Difference	Abs Val
Player 1	233	−38	38	305	+34	
Player 2	259			392		121
Player 3	250	−21		197	−74	
Player 4	244		27	208		
Player 5	286	+15		222	−49	49
Total	1,272		113	1,324	−31	
Average		−16.6				

5. Study the numbers in the table. Which team do you think did a better job of guessing the number of jelly beans? Explain your answer.

_____

_____

_____

_____

6. You could take a wild guess to try to get the number of jelly beans, or you could use math skills to try to get closer to the real number. What math strategy could you use? Explain how you would use this strategy to find the number of jelly beans.

_____

_____

_____

_____

7. Another game is "Factor Toss." There are 9 plastic whole numbers, 1–9, and a frame with different sections as shown below. You try to toss the numbers into the right section. The area where ovals overlap is for factors of two or three numbers. Put the numbers 1–9 in the right sections. The number 5 is placed for you.

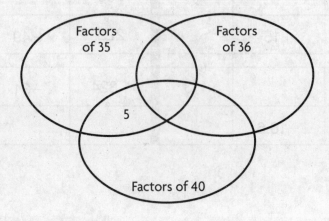

1. Choose the expressions that make the equation true. Mark all that apply.

   $$\frac{8}{15} \div \frac{4}{5} = \underline{\hspace{2cm}}$$

   (A) $\frac{1}{3}$

   (B) $\frac{2}{3}$

   (C) $\frac{20}{60}$

   (D) $\frac{40}{60}$

2. Select the expressions that are equivalent to 81. Mark all that apply.

   (A) 3^4

   (B) 9^9

   (C) $3^3 \times 3$

   (D) $3 \times 3 \times 9$

3. Fernando donates $2 to a local charity organization for every $15 he earns. Cleo donates $4 for every $17 she earns. Is Fernando's ratio of money donated to money earned equivalent to Cleo's ratio of money donated to money earned? Explain.

   _____

   _____

4. Charlotte has 2 oranges and 5 apples. Select the ratios that compare the number of oranges to the total number of fruits. Mark all that apply.

   (A) 2 to 5

   (B) 2 to 7

   (C) 2 : 7

   (D) 5 : 2

   (E) $\frac{2}{5}$

   (F) $\frac{2}{7}$

GO ON

5. Joan can buy 20 cans of cat food for $14 at the grocery store. She can buy 25 cans for $15 at the pet store. Which store sells cat food for less per can? Use numbers and words to explain your answer.

6. Select the situations that can be represented by a negative number. Mark all that apply.

Ⓐ The city of New Orleans is 7 feet below sea level.

Ⓑ The average temperature in San Diego in July is 70°F.

Ⓒ Dana scored 40 points playing a video game.

Ⓓ Jake withdraws $40 from the bank.

7. Select the numbers that would be located between ⁻1 and ⁻2 on a number line. Mark all that apply.

Ⓐ $1\frac{1}{5}$

Ⓓ $^{-}1\frac{4}{9}$

Ⓑ ⁻1.8

Ⓔ ⁻1.09

Ⓒ 0.01

8. Which expresses the calculation *multiply 7 and p, and then subtract 28*?

Ⓐ $p - 28$

Ⓒ $7p - 28$

Ⓑ $7p + 28$

Ⓓ $7(p - 28)$

GO ON ➡

9. Carrie is 3 years older than her sister Lucy. Let c represent Carrie's age. Identify the expression that can be used to find Lucy's age.

Ⓐ $c - 3$

Ⓑ $c + 3$

Ⓒ $3c$

Ⓓ $\dfrac{c}{3}$

10. Identify the quadrant where each point is located. Write each point in the correct box.

$(^-2, ^-4)$ $(4, 1)$ $(^-4, 8)$

$(9, ^-5)$ $(3, ^-7)$ $(^-1, 5)$

Quadrant I	Quadrant II	Quadrant III	Quadrant IV

11. Elena ran 15 miles in 5 days. She runs the same distance each day. Write and solve an equation to find how far Elena runs each day.

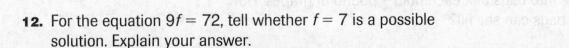

12. For the equation $9f = 72$, tell whether $f = 7$ is a possible solution. Explain your answer.

GO ON

13. Jamie earns $12 per hour for babysitting. Write a linear equation that gives the wages, *w*, in dollars that Miranda earns in *h* hours.

_____

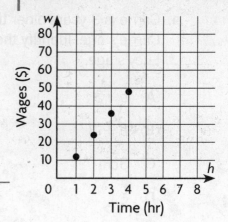

14. Select the expressions that are equivalent to $\frac{1}{8}$. Mark all that apply.

Ⓐ $\left(\frac{1}{2}\right)^3$

Ⓑ $\left(\frac{1}{3}\right)^2$

Ⓒ $(0.5)^3$

Ⓓ $(0.125)^3$

15. Laura collected 12 more autographs than Lance. Let *a* represent the number of Lance's autographs. Identify the expression that can be used to find the number of autographs that Laura collected.

Ⓐ $12a$ Ⓒ $12a + 2$

Ⓑ $a - 12$ Ⓓ $a + 12$

16. Jessica has $2\frac{1}{2}$ pounds of grapes. She wants to divide the grapes into bags that each hold $\frac{1}{4}$ pound of grapes. How many bags can she fill?

_____ bags

GO ON

17. To reach her summer reading goal, Elisa has to read at least 30 books. An inequality $b \geq 30$ represents the number of books she needs to read. Which of the following are solutions to this inequality? Mark all that apply.

Ⓐ 28   Ⓒ 30

Ⓑ 29   Ⓓ 31

18. Jake wants to rent a boat to get around the lake. A canoe rental costs $6 for 30 minutes. A rowboat rental costs $8 for 45 minutes. Which of these will cost less to rent for 90 minutes? Complete the tables of equivalent ratios to support your answer.

Canoe				
Cost (dollars)	6			
Time (minutes)	30		90	120

Rowboat				
Cost (dollars)	8			32
Time (minutes)	45	90		

19. How many $\frac{7}{8}$-ounce servings of cereal are there in a $12\frac{1}{4}$-ounce box?

_____ servings

20. While playing a video game, Ruth earned 5 base points, *b*; 9 bonus points, *x*; and 4 advanced points, *a*. Write an algebraic expression for the total number of points Ruth scored.

_____

21. A membership at a gym costs $45 per month. Select the gyms that offer memberships at a lower unit rate. Mark all that apply.

(A) Gym A: $90 for 2 months

(B) Gym B: $120 for 3 months

(C) Gym C: $172 for 4 months

(D) Gym D: $184 for 4 months

22. Mrs. Romero needs at least 12 students to volunteer to help out with the school play. The inequality $v \geq 12$ represents the number of students, *v*, who must volunteer. Select possible solutions of the inequality. Mark all that apply.

(A) 8

(B) 10

(C) 12

(D) 13

(E) 25

(F) 30

23. A cube has side lengths that measure 6 units each. Use the formula $V = s^3$ for volume. Which shows the volume? Mark all that apply.

(A) 2 (6 units × 6 units)   (C) 216 cubic units

(B) 3 × 6 units   (D) 6^3 cubic units

Name _____

Middle-of-Year Test
Page 7

24. Choose the word that makes the statement correct.

If both the *x*- and *y*-coordinates are

| positive |
| negative |
| equal |

, the point is

always to the right of the *y*-axis and above the *x*-axis.

25. A store is offering a special deal. Shirts cost $18 each and the customer can get an unlimited number of shirts delivered for $10. Write an expression that can be used to find the cost of any number of shirts and the delivery charge.

_____

26. To rent a bicycle, there is an insurance fee of $14. Then the rent is $6 per day. Use the equation $c = 6d + 14$ to complete the table.

Input	Output
Days, *d*	Cost ($), *c*
2	
4	
6	
8	

GO ON

27. Which expression can be used to represent "2 times more than the sum of a number and 27.9"?

(A) $2n + 27.9$

(B) $n \times 2 \times 27.9$

(C) $2(n + 27.9)$

(D) $2n(2 \times 27.9)$

28. What is the value of the expression $12 - 6f + 2f$ when $f = 0.5$?

(A) 2

(B) 8

(C) 10

(D) 12

29. Which points are on the same side of 0 on a number line as the opposite of $-\frac{3}{8}$? Mark all that apply.

(A) $\frac{3}{4}$

(B) $-\frac{5}{8}$

(C) -2

(D) 0.5

(E) 4.3

GO ON

Cooperstown Bound

Mr. and Mrs. Isaac and their three children take a trip by car to Cooperstown.

1. The Isaacs live *m* miles from Cooperstown, and they drive 30 miles while they are in the town. Write an algebraic expression using *m* to show how many miles they drive from the time they leave home until they return. Explain your answer.

2. The total distance the Isaacs travel is 450 miles. Write and solve an equation to find *m*, the distance from their home to Cooperstown.

3. The Isaacs' car can go 30 miles on 1 gallon of gas. Write and solve an equation to find *g*, the number of gallons of gas they use driving 450 miles.

GO ON

4. On the interstate highway the Isaacs can drive at a steady
65 miles per hour. Fill in the table to show how far they travel
in 1, 2, and 3 hours.

Time *t* in hours	Distance *d* in miles
0	0
1	
2	
3	

5. Make a graph using the numbers in your table, with time
t on the *x*-axis and distance *d* on the *y*-axis.

6. Write an equation to tell how far the Isaacs travel, *d*, in *t* hours on the interstate.

Use your equation to figure out how far they go in $1\frac{1}{2}$ hours.

7. The Isaacs have budgeted *x* dollars per day for each of their 3 days at a hotel and *y* dollars per day for food for each of the 3 days. Write two different expressions to give the total for hotel and food. Describe your expressions.

8. One of the highlights of their trip is a visit to the Baseball Hall of Fame. Let *a* = the cost of an adult ticket for the Hall of Fame. Let *c* = the cost of a child's ticket. Write an expression using *a* and *c* to tell how much it costs for all 5 people in the Isaac family to get tickets for the Hall of Fame. Explain your answer.

The adult tickets cost $19.50 each, and the children's tickets cost $7.00 each. Substitute these values into your expression and evaluate the expression to find the amount they pay for all the tickets.

GO ON

9. The Baseball Hall of Fame opened in 1939. Write and solve an equation to tell a, how old the Hall of Fame is. Explain your answer.

10. They also visit the Farmers' Museum. One of the exhibits at the Farmers' Museum is the "Cardiff Giant," a huge carved stone figure that was buried on a farm in New York in the 1860s as a practical joke and money-making scheme. George Hull paid $2,600 to have the figure carved and transported, and then he made $30,000 in a short time from people paying 50¢ each to see the giant. Write and solve an equation to find n, how many people came to see the giant during this time.

11. Tickets for the Farmers' Museum are $12 for adults and $6 for children. A ticket for *both* the Farmers' Museum and the Baseball Hall of Fame is x dollars for an adult and y dollars for a child. Write an expression to show how much the Isaac family can save by buying double tickets for everyone rather than individual tickets. (Also use the information in question 8.) You do not need to simplify the expression.

12. The Isaacs budget $30 for souvenirs. Let s stand for the cost of souvenirs. The sales tax is 8% of the cost, so the total for the souvenirs is $s + 0.08s$. If each Isaac child chooses a souvenir costing $9.25, will they be able to keep their souvenir purchase within their budget?

1. There were $14\frac{1}{4}$ cups of apple juice in a container. Each day, Elise drank $1\frac{1}{2}$ cups of apple juice. Today, there is $\frac{3}{4}$ cup of apple juice left.

   For how many days did Elise drink $1\frac{1}{2}$ cups of apple juice?

   _____ days

2. Use exponents to write the expression.

   $2 \times 2 \times 2 \times 4 \times 4$

   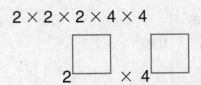

   $2^{\square} \times 4^{\square}$

3. The Tuckers drive at a rate of 20 miles per hour through the mountains. Use the ordered pairs to graph the distance traveled over time.

Distance (miles)	20	40	60	80	100
Time (hours)	1	2	3	4	5

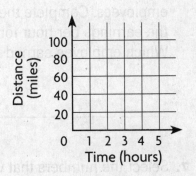

4. Ms. Wilson gave a quiz to her science class. The students' scores are listed in the table. Make a dot plot of the data.

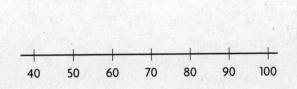

Science Test Scores				
90	90	50	70	70
80	90	50	70	60
90	70	60	80	100
70	50	80	90	90
80	70	80	100	70

GO ON

5. The data set shows the scores of three players for a board game.

Board Game Scores			
Player A	90	90	90
Player B	110	100	90
Player C	95	100	95

Choose the word that makes the statement correct.

The mean absolute deviation of Player B is | greater than / less than / equal to | the

mean absolute deviation of Player C.

6. The table shows the total earnings and the number of hours worked for three employees. Complete the table by finding the earnings per hour for each employee. Which employee earned the least per hour?

_____

Employee	Total Earned (in dollars)	Number of Hours Worked	Earnings per Hour (in dollars)
1	$34.02	3.5	
2	$42.75	4.5	
3	$52.65	5.4	

7. Select the numbers that would be located between 0 and ⁻1 on a number line. Mark all that apply.

Ⓐ $\frac{3}{5}$

Ⓓ $\frac{-6}{8}$

Ⓑ $^-1\frac{1}{2}$

Ⓔ 0.11

Ⓒ ⁻0.01

8. Which expression is equivalent to $6(x - 4)$?

Ⓐ $6x - 10$

Ⓒ $6x - 24$

Ⓑ $x - 10$

Ⓓ $x - 24$

GO ON

9. The number of cups of hot chocolate sold during lunch time at five different snack carts are 8, 16, 18, 19, and 24. The mean number of cups of hot chocolate sold is 17. Identify the outlier. If the outlier increased, would that affect the mean of this set of data? Explain your answer.

_____

_____

10. Identify the quadrant where each point is located. Write each point in the correct box.

(7, 6)   (⁻5, 9)   (8, 4)

(⁻3, ⁻1)   (6, ⁻2)   (4, ⁻2)

Quadrant I	Quadrant II	Quadrant III	Quadrant IV

 11. Melinda drives 120 miles in 2 hours at a constant speed. If she keeps driving at a constant speed, how far will Melinda drive in 5 hours?

Ⓐ 180 miles   Ⓓ 360 miles

Ⓑ 240 miles   Ⓔ 600 miles

Ⓒ 300 miles

12. Select the pairs of points that have a distance of 2 units between them. Mark all that apply.

Ⓐ (⁻4, 4) and (⁻4, 2)   Ⓒ (⁻3, 3) and (⁻3, 1)

Ⓑ (1, 1) and (1, ⁻3)   Ⓓ (1, ⁻6) and (1, ⁻10)

GO ON

13. Circle <, >, or =.

13a. ⁻12 [< > =] ⁻9   13c. ⁻1.5 [< > =] ⁻2

13b. ⁻8 [< > =] ⁻10   13d. ⁻0.6 [< > =] ⁻1.6

14. Match the inequality to the word sentence it represents.

$a <$ [80] •   • No more than 80 passengers are allowed to ride on the bus at the same time.

$b \leq$ [80] •   • Mikayla saved less than $80.

$c >$ [80] •   • Ava spent more than $80 on concert tickets.

15. Identify the expression that can be used to express the calculation *subtract 17 from q.*

(A) $q - 17$ (C) $q + 17$

(B) $17q$ (D) $17q + 17$

16. Sarabeth asks the timekeeper at a car race, "What was the fastest speed reached by the winner of the race?" Explain why this is not a statistical question.

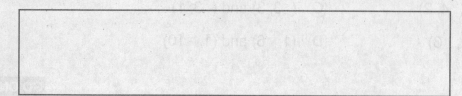

GO ON ▶

17. Maria needs to save at least $130 to buy a new bike. The inequality $x \geq \$130$ represents the amount of money she needs to save. Which of the following are solutions to this inequality? Select all that apply.

    (A) $135   (C) $125

    (B) $130   (D) $120

18. Jordan surveyed a group of randomly selected smartphone users and asked them how many applications they have downloaded onto their phones. The dot plot shows the results of Jordan's survey. Select the statements that describe patterns in the data. Mark all that apply.

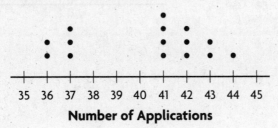

Number of Applications

    (A) The modes are 37 and 42.

    (B) There is a gap from 38 to 40.

    (C) There is a cluster from 41 to 44.

    (D) There is a cluster from 35 to 36.

19. Write the values in order from greatest to least.

_____

GO ON ▶

20. Javier bought *x* pounds of olives at $4 per pound and
y pounds of feta cheese at $2 per pound. Write an algebraic
expression for the cost of Javier's purchase.

_____

21. A teacher surveys his students to find out how much time the
students spent using social media over the weekend.

He uses
| hours |
| minutes |
| seconds |
as the unit of measure.

Time Spent Online (hours)			
2.5	2	5	3.5
1.5	3	3	4.5
3.5	4	4.5	5

He made
| 11 |
| 12 |
| 13 |
observations.

22. Draw a line to match each solid figure with its net.

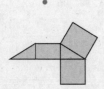

23. A rectangular prism measures 6 units long, 3 units wide, and
2 units high. Select the expressions that use the formula for
volume, $V = Bh$, to show the volume of the rectangular prism.
Mark all that apply.

Ⓐ $2(6 \times 3 \times 2)$ Ⓒ $6 \times 3 \times 2$

Ⓑ $6 + (3 \times 2)$ Ⓓ $2(6 \times 3)$

24. Explain how to graph points *A* (4, 0), *B* (4, ⁻3), and *C* (0, ⁻3) on the coordinate plane. Then, explain how to graph point *D* so that *ABCD* is a rectangle.

25. Name the regular polygon and find its area. Show your work.

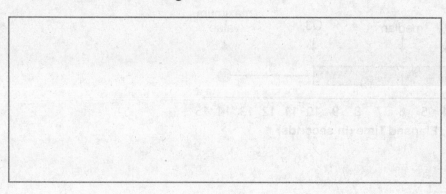

4.8 in.

4 in.

_____

_____

26. A line segment has endpoints (3, 14) and (3, 22). Which of the following statements is true?

Ⓐ The segment has a length of 8 units and is horizontal.

Ⓑ The segment has a length of 8 units and is vertical.

Ⓒ The segment has a length of 18 units and is horizontal.

Ⓓ The segment has a length of 18 units and is vertical.

GO ON ➡

27. Use the box plot to calculate the range and interquartile range for the data displayed.

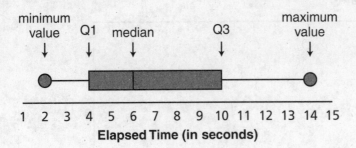

Range: _____

Interquartile range: _____

28. Lisa has a keepsake box that is in the shape of a rectangular prism.

The volume is

$3,937\frac{1}{2}$ in.³

$2,150\frac{1}{3}$ in.³

$1,230\frac{1}{2}$ in.³

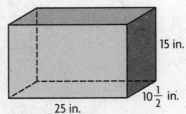

29. Jamie said |⁻2| equals |2|. Is Jamie correct? Draw a number line and use words to support your answer.

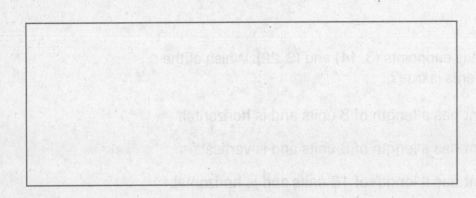

GO ON

Trip to Mexico

Denine took a trip to Mexico. While she was there she had to use some of her math skills to understand distances and other units of measure and to exchange money.

1. In Mexico, people use pesos for money. There are about 13 pesos in 1 U.S. dollar.

   a. About how much is 1 peso worth in dollars? Show your work and give your answer to the nearest hundredth of a dollar and the nearest cent.

   b. Denine sees a scarf for 70 pesos and wants to know if it is in her budget for souvenirs. Use ratio reasoning to find the approximate cost of the scarf in dollars. Round your answer to the nearest half-dollar. (There will be a small error from rounding the value of the peso.)

2. Denine goes on a tour to see the Pyramid of the Sun and the Pyramid of the Moon. There are 29 men and 24 women on the tour. Write the ratio that compares the number of men on the tour to the total number of people on the tour.

3. Denine reads that the Pyramid of the Sun is more than 71 meters tall. One meter is about 3 feet. About how tall is the pyramid in feet?

4. When Denine gets home, she wants to make a model of the Pyramid of the Sun. The base of the real pyramid is about 224 meters wide on each side. If she makes her model 24 inches wide at the base, about how tall will the model be? Use ratio reasoning to find the answer.

GO ON

5. Denine's hotel room costs 716 dollars for 4 nights. Her friend
   Claudia is staying at another hotel. Claudia's hotel room costs
   495 dollars for 3 nights. Whose hotel room costs less per night?

6. At a market, Denine buys a bag of 6 mangoes for 15 pesos.
   What is the unit price for 1 mango?

7. One market sells a 3-kilogram bag of tortillas for 42 pesos, and
   another sells 2 kilograms of tortillas for 26 pesos. Which unit
   price is lower?

8. Denine is making a vegetable soup called gazpacho. The recipe asks for 680 milliliters of tomato juice. One ounce is about 30 milliliters. About how many ounces of tomato juice does Denine need?

 9. Denine learns that about 3 out of 10 people in Mexico are under the age of 15. What percent of the population is under the age of 15?

 10. If the population of Mexico is about 116 million, approximately how many children under the age of 15 are in the country?

Beaded Baskets

Mr. Appleton teaches an arts and crafts class at his local community center. Last month he taught his students how to make beaded baskets.

Mr. Appleton distributed a basic bead pattern to his students. The pattern shows that the base of the basket is made up of 48 beads surrounding a plastic disc. The entire basket is made from 2 colors of beads.

Mr. Appleton adjusted the ratio of beads in the pattern, resulting in two additional design choices. The new designs are similar to the original in that they also have a base made up of 48 beads surrounding a plastic disc. The students were instructed to select one of the designs.

Basket A: 2 colors of beads at a ratio of 2:1

Basket B: 2 colors of beads at a ratio of 3:1

Basket C: 2 colors of beads at a ratio of 3:5

GO ON

Students also created nametags to display along with their
baskets.

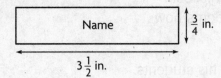

Mr. Appleton recorded information about each student's basket
in a chart.

He wanted to use the information for future beading projects.

Name	Basket Design	Height of Basket (inches)	Number of Beads Used
Ava	A	$3\frac{1}{2}$	288
Viola	A	4	360
Martin	C	$3\frac{1}{2}$	288
Joe	B	$4\frac{3}{4}$	432
Claire	B	$4\frac{3}{4}$	432
Evan	B	$4\frac{3}{4}$	460
Sal	A	$3\frac{3}{4}$	330
Allie	C	$5\frac{1}{4}$	576
Jose	B	4	360
Kim	B	$5\frac{1}{4}$	480

GO ON ➡

1. Ava chose to make her basket using red and gold beads. She created the base of the basket using 32 red beads and 16 gold beads. Explain how to determine which of the baskets A, B, or C Ava made.

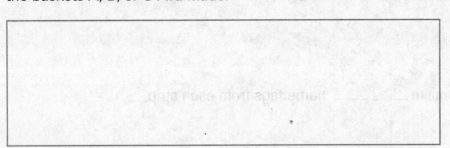

2. Both Martin and Kim used green and yellow beads to make their baskets. Martin chose to make Basket C, while Kim chose to make Basket B. Kim says that they should both use the same number of green beads to make up the base. Is Kim correct? Explain your answer.

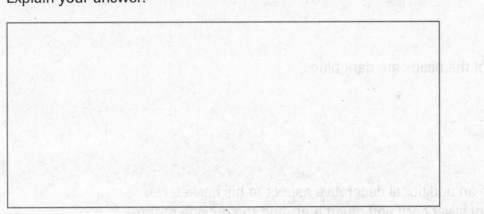

3. Sal enjoyed the beading project so much that he decided to make a bead basket at home. Sal sees a sign in a craft store advertising a pack of 50 crystal beads for $3. He sees the same type of beads advertised online for $4 for a pack of 80 beads. Which is the better buy? Explain your answer.

4. Mr. Appleton bought strips of oak tag from which to make the nametags for the basket display. Each strip of oak tag measures 6 inches long. How many $\frac{3}{4}$-inch wide name tags can he cut from each strip of oak tag?

Mr. Appleton can make _____ name tags from each strip.

5. Four-tenths of the beads in Mr. Appleton's collection are blue. One-half of those blue beads are dark blue. What fraction of all the beads in Mr. Appleton's collection are dark blue?

_____ of the beads are dark blue.

6. Kim included an additional decorative aspect to his basket. He cut a length of silver cord and glued it around the outside of the basket. Kim used the cord to divide the basket into equal parts. What is the height of the basket above the cord?

The height of the basket is _____ inches.

GO ON ➡

7. Write an expression that describes the beads in Allie's
pattern for the base of her basket. Evaluate the expression.
Use numbers and words to explain your answer. Then, draw
a picture of a string of beads that reflects the expression.

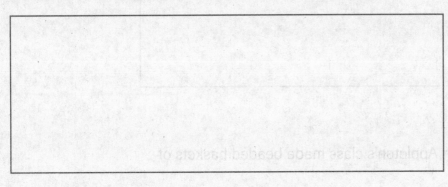

 8. Mr. Appleton needs to return one of the boxes of beads
he purchased. He found that $\frac{2}{5}$ of the beads, or 200 beads,
were chipped or defective in some way. Write an equation to
show how many beads were in the box. Then, use words and
numbers to explain how to solve the equation.

9. Find the mean number of beads used by students. Then find the absolute mean deviation. Show your work.

10. The students in Mr. Appleton's class made beaded baskets of various heights.

10a. Draw a dot plot to show the heights of the students' beaded baskets.

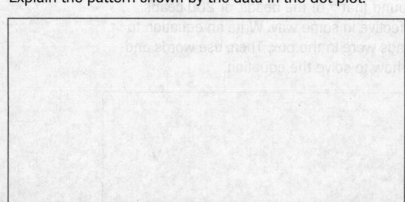

10b. Describe the data in the dot plot you created for 10a. Explain the pattern shown by the data in the dot plot.